edexcel
advancing learning, changing lives

Edexcel AS Biology Revision Guide

for SNAB and concept-led approaches

REVISION GUIDE

A PEARSON COMPANY

Published by Pearson Education Limited, a company incorporated in England and Wales, having its registered office at Edinburgh Gate, Harlow, Essex, CM20 2JE. Registered company number: 872828

Edexcel is a registered trade mark of Edexcel Limited

Text © Pearson Education Limited 2009

All past exam questions © Edexcel

The rights of Gary Skinner, Stephen Winrow-Campbell and John Dunkerton to be identified as the authors of this work have been asserted by them in accordance with the Copyright, Designs and Patents Act of 1988.

First published 2009

10 9 8 7 6

British Library Cataloguing in Publication Data
A catalogue record for this book is available from the British Library

ISBN 978 1 846905 98 8

External project management by Sue Kearsey
Edited by Liz Jones
Typeset by 320 Design Ltd
Illustrated by Oxford Designers & Illustrators
Cover photo © Jupiter Unlimited
Printed in Malaysia, KHL-CTP

Acknowledgements
Edexcel review by Martin Furness-Smith, UYSEG review by Anne Scott
We would like to thank Damian Riddle, Anne Scott and Elizabeth Swinbank for their contributions

Disclaimer
This material has been published on behalf of Edexcel and offers high-quality support for the delivery of Edexcel qualifications.

This does not mean that the material is essential to achieve any Edexcel qualification, nor does it mean that it is the only suitable material available to support any Edexcel qualification. Edexcel material will not be used verbatim in setting any Edexcel examination or assessment. Any resource lists produced by Edexcel shall include this and other appropriate resources.

Copies of official specifications for all Edexcel qualifications may be found on the Edexcel website: www.edexcel.com

Contents

Revision techniques

Getting started can be the hardest part of revision, but don't leave it too late. Revise little and often! Don't spend too long on any one section, but revisit it several times, and if there is something you don't understand, ask your teacher for help.

Just reading through your notes is not enough. Take an active approach using some of the revision techniques suggested below.

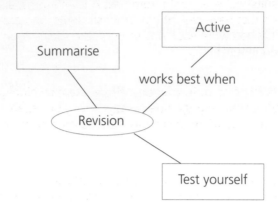

Summarising key ideas

Make sure you don't end up just copying out your notes in full. Use some of these techniques to produce condensed notes.

- Tables and lists to present information concisely
- Index cards to record the most important points for each section
- Flow charts to identify steps in a process
- Diagrams to present information visually
- Spider diagrams, mind maps and concept maps to show the links between ideas
- Mnemonics to help you remember lists
- Glossaries to make sure you know clear definitions of key terms

Include page references to your notes or textbook. Use colour and highlighting to pick out key terms.

Active techniques

Using a variety of approaches will prevent your revision becoming boring and will make more of the ideas stick. Here are some methods to try.

- Explain ideas to a partner and ask each other questions.
- Make a podcast and play it back to yourself.
- Use PowerPoint to make interactive notes and tests.
- Search the internet for animations, tests and tutorials that you can use.
- Work in a group to create and use games and quizzes.

Test yourself

Once you have revised a topic, you need to check that you can remember and apply what you have learnt.

- Use the questions from your textbook and this revision guide.
- Get someone to test you on key points.
- Try some past exam questions.

Check the spec.
If you use resources from elsewhere, make sure they cover the right content at the right level.

How to use this Revision Guide

Welcome to your **Edexcel AS Biology Revision Guide**, perfect whether you're studying Salters Nuffield Advanced Biology (the orange book), or the 'concept-led' approach to Edexcel Biology (the green book).

This unique guide provides you with tailored support, written by senior examiners. They draw on real 'ResultsPlus' exam data from past A-level exams, and use this to identify common pitfalls that have caught out other students, and areas on which to focus your revision. As you work your way through the topics, look out for the following features throughout the text.

ResultsPlus Examiner's Tip
These sections help you perform to your best in the exam by highlighting key terms and information, analysing the questions you may be asked, and showing how to approach answering them. All of this is based on data from real-life A-level students!

ResultsPlus Watch Out
The examiners have looked back at data from previous exams to find the common pitfalls and mistakes made by students – and guide you on how to avoid repeating them in *your* exam.

Quick Questions
Use these questions as a quick recap to test your knowledge as you progress.

Thinking Task
These sections provide further research or analysis tasks to develop your understanding and help you revise.

Worked Example
The examiners guide you through complex equations and concepts, providing step-by-step guidance on how to tackle exam questions.

Each topic also ends with:

Topic Checklist
This summarises what you should know for this topic, which specification point each checkpoint covers and where in the guide you can revise it. Use it to record your progress as you revise.

ResultsPlus Build Better Answers
Here you will find sample exam questions with exemplar answers, examiner tips and a commentary comparing both a basic and an excellent response; so you can see how to get the highest marks.

Practice Questions
Exam-style questions, including multiple-choice, offer plenty of practice ahead of the exam.

Both Unit 1 and Unit 2 conclude with a **Specimen Paper** to test your learning. These are not intended as timed, full-length papers, but provide a range of exam-style practice questions covering the range of content likely to be encountered within the exam.

The final unit consists of advice and support on how to choose a good issue to research or place to visit, and how to get the best possible marks for the report that you write on your findings.

Answers to all the in-text questions, as well as detailed, mark-by-mark answers to the specimen papers, can be found at the back of the book.

We hope you find this guide invaluable. Best of luck!

Question types in GCE Biology

Multiple choice

A good multiple choice question gives you the correct answer and other possible answers which seem plausible.

Triglycerides are composed of: 3 glycerol molecules and 3 fatty acid molecules ☐ 1 glycerol molecule and 3 fatty acid molecules ☐ 1 glycerol molecule and 1 fatty acid molecule ☐ 3 glycerol molecules and 1 fatty acid molecule ☐	(1)

The best way to answer a multiple choice question is to read the question and try to answer it before looking at the possible answers. If the answer you thought of is amongst the possible answers – job done! Just have a look at the other possibilities to convince yourself that you were right.

If the answer you thought of isn't there, look at the possible answers and try to eliminate wrong answers until you are left with the correct one.

You don't lose any marks by having a guess (if you can't work out the answer) – remember you won't score anything by leaving the answer blank! If you narrow down the number of possible answers, the chances of having a lucky guess at the right answer will increase.

To indicate the correct answer, put a cross in the box following the correct statement. If you change your mind, put a line through the cross and fill in your new answer with a cross.

How Science Works

The idea behind How Science Works is to give you insight into the ways in which scientists work: how an experiment is designed, how theories and models are put together, how data are analysed, how scientists respond to factors such as ethics, and so on, and the way society is involved in making decisions about science.

Many of the HSW criteria are practical and will be tested as part of your practical work. However, there will be questions on the written unit papers that cover some HSW criteria. Some of these questions will involve data or graph interpretation (HSW 5) – see the next section.

The other common type of HSW question will be based on the core practicals. Questions will concentrate not so much on what you did, but why various steps in the core practical were important. It's important, therefore, that you know what the various steps in each core practical were designed to do; and that you revise the core practicals.

For example, think about the questions that could be asked in a Unit 2 paper on the core practical in Topic 3: 'Describe the stages of mitosis and how to prepare and stain a root tip squash in order to observe them practically.' Here, suitable questions could include:
- Why do we use only the tip of the root?
- Which stain do we use?
- Why do we place the cut tip in acid before staining?
- What safety precautions would be relevant here, and why?

You'll also commonly get asked questions involving designing an investigation: these are likely to involve pieces of familiar practical work. The **CORMS** prompt may be useful here:

Control – Are you investigating simply with / without a particular factor?
What range of values are you looking at?
Organism – Are you using organisms of the same sex / age / size / species?
Repeat – Take readings more than once and average
Measure – What are you measuring? How will you measure this? What units?
Same – Which variable(s) are you keeping constant?

Other HSW questions may concentrate on ethical issues surrounding topics such as gene therapy.

Interpretation of graphs

The graph below shows the results of a survey in America, on the incidence of heart disease in adults aged 18 and older.

> Using the information in the graph, describe how the incidence of heart disease is affected by age and gender. (3)

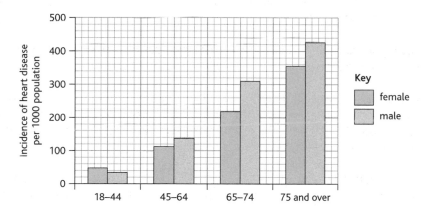

You almost always get one mark for stating the general trend – in this case that the incidence of heart rate increases, with increasing age, for both genders. You can then concentrate on individual aspects of the data. In this case, what stands out most is that females have a lower incidence of heart disease than males, except in the 18–44 age group. You may also comment that the difference between the incidence in males and females is largest in the 65–74 age group.

Finally, there is always one mark for manipulation of data. Note that this must be *manipulation* – you don't get marks for reading off the graph and stating the numbers, you have to do something with them!

Extended questions

In the A2 units (Unit 4 and Unit 5) you will come across questions with larger numbers of marks, perhaps up to 6 or 7 marks in the question.

Questions in these units are designed to be synoptic – in other words, they are designed for you to show knowledge gained in earlier units. Bear this in mind when you answer the question: try to include relevant knowledge from your AS course when answering these questions.

Remember, too, that if the question is worth 6 marks, you need to make 6 creditworthy points. Think about the points that you will make and put them together in a logical sequence when you write your answer. On longer questions, the examiners will be looking at your QWC (Quality of Written Communication) as well as the answer you give.

Transport and circulation

Water

Water is the medium of transport in all living things because it is an excellent **solvent**. This solvent ability comes from the fact that water molecules have uneven charge distribution, one end of the molecule is slightly positive δ^+, the other end is slightly negative δ^-. The water molecule is said to be polar.

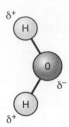

The δ^+ and δ^- denote small amounts of negative and positive charge on the water molecule.

- Ionic substances such as sodium chloride dissolve easily in water because the positively and negatively charged ions are separated due to the dipole nature of water.

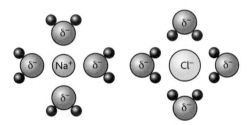

Sodium (+ charge) and chloride (– charge) ions surrounded by water molecules, i.e. dissolved.

- Molecules, such as glucose and amino acids, are also polar due to charged groups (in sugars it is –OH, in amino acids –NH$_2$ and –COOH) and therefore dissolve easily.

- Hydrogen bonds (H bonds) form between molecules due to their dipole nature, holding the molecules together. So water is a liquid at room temperature, unlike other small molecules such as CO_2, which is a gas.

H bonding creates **cohesion**, important in transport of water and dissolved substances in the xylem (page 78).

H bonding means it takes a lot of energy to warm water up (high specific heat capacity) and also to turn it into a gas (high latent heat of vaporisation). So temperature fluctuations in living things are small. Also sweating and transpiration take energy from the body and cool it down.

Three water molecules held together by hydrogen bonds.

Why do we need to transport?

Organisms need substances to enter and leave the body in sufficient quantities. In very small organisms simple diffusion is fast enough to supply their needs. Big organisms have a much smaller surface area to volume ratio (SA/V). So there is not enough surface to serve the needs of the large volume inside by diffusion.

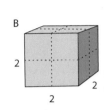
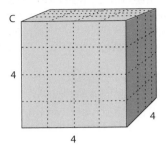

Cube A has a volume of 1 cm³ and an area of 6 cm². This gives it a SA/V ratio of 6/1, i.e. 6. Think about the SA/V ratio for cubes B and C, and then look at the thinking task below.

To overcome the limitations of diffusion, large organisms have special organs, like lungs (Topic 2), to increase the surface area for exchange. A **mass transport system** (e.g. heart and circulation) moves the exchanged materials around the body.

The structure and function of arteries, veins and capillaries

Vessel	Structure	Functional significance
artery	relatively thick wall smooth muscle	• withstands high blood pressure • alters diameter of lumen to vary blood flow
	elastic fibres	• allow walls to stretch when blood is pumped into the artery and then recoil, smoothing blood flow
	lined with smooth layer of endothelial cells narrow lumen	• low friction surface to ease blood flow
capillaries	very thin wall (just one cell thick)	• allows rapid exchange between blood and tissues
veins	relatively thin wall very little smooth muscle or elastic fibres wide lumen valves	• blood under low pressure • no pulse of blood so no stretching and recoiling • large volume acts as blood reservoir • stop backflow

Quick Questions

Q1 List *three* ways in which hydrogen bonds are important in giving water its properties.

Q2 Which part of a water molecule is slightly negative and which is slightly positive?

Q3 Put these organisms in order of likely SA/V ratio. Show clearly which is highest and which is lowest.
yeast, golden retriever, goldfish, killer whale, elephant, *Daphnia*

Thinking Task

Q1 Calculate the SA/V ratio of cubes B and C in the diagram above. Relate your findings to the limitations of diffusion in large organisms.

Cardiac cycle and heart rate

The heart pumps the blood around the body. It is difficult to study the human heart because it is deep within the body. However, we can study the functioning of the heart in an organism in which the heart is visible without the need for damaging and invasive procedures, such as the water flea, *Daphnia* sp. This invertebrate is suitable for this experiment because it:

- is abundant and easily obtained
- is transparent
- has a very simple nervous system and will not suffer 'stress' such as a mammal might, which makes it ethically more appropriate.

Many substances, such as hormones and dietary components, can affect heart rate. A core practical is to look at the effect of caffeine.

There are two ways to do this experiment:

- Put *Daphnia* in a caffeine solution and compare heart rates with control *Daphnia* in plain pond water.
- Put *Daphnia* in a range of different strength caffeine solutions.

Making this experiment successful

• immobilise the *Daphnia*	Use strands of cotton wool in a small dish of the experimental solution.
• control other variables, such as water temperature, *Daphnia* size, etc.	Difficult to maintain a constant temperature but should be monitored with a thermometer in the water. *Daphnia* of similar size, etc. should be used for all measurements.
• *accurate* measurement of heart rate	A dot is put on a piece of paper (in an S shape to avoid putting one dot on top of another) or clicking a button on a calculator.
• repeatability	Ensure that variables other than caffeine concentration are controlled.

Typical student data from this experiment:

	Mean heart rate/beats min⁻¹	
Daphnia	**With 0.5% caffeine**	**Plain filtered pond water (0% caffeine)**
1	170	140
2	164	182
3	170	162
4	186	194
5	176	180

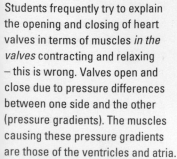
The cardiac cycle

The human heart is basically two muscular pumps, one sending the blood to the lungs for removal of carbon dioxide and oxygenation, and the other sending the blood round the body. A system of valves ensures that blood flows through the heart in one direction.

The atria and ventricles contract during systole and relax during **diastole**. Elastic recoil of the heart muscles lowers pressure in the atria and ventricles. Blood is drawn into the heart from the arteries and veins, closing the semi-lunar valves.

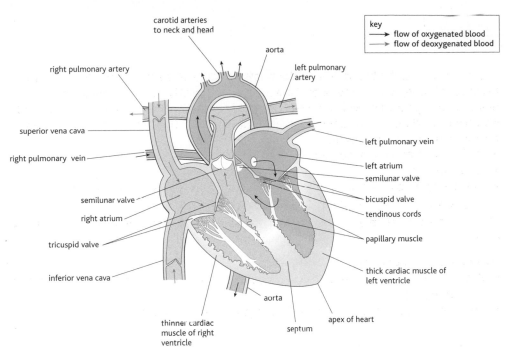

Vertical section through the heart to show blood flow.

This is the sequence of events on the *left side* of the heart during a single 'beat':

Blood drains into the left atrium from lungs along the pulmonary vein

↓

Raising of the blood pressure in the left atrium forces the left atrio-ventricular valve open

↓

Contraction of the left atrial muscle (left atrial systole) forces more blood through the valve

↓

As soon as left atrial systole is over, the left ventricular muscles start to contract. This is called left ventricular systole. This forces the left A–V valve closed and opens the valve in the mouth of the aorta (semilunar valve). Blood then leaves the left ventricle along the aorta.

On the right side, at the same time, blood enters from the body along the vena cava. The right atrial muscle contracts and the right AV valve opens due to the pressure difference. The blood enters the right ventricle. Now the right ventricle muscle contracts, forcing the semilunar valve in the mouth of the pulmonary artery open and the blood leaves through this valve along the artery to the lungs.

The atria and ventricles contract during systole and relax during **diastole**. Elastic recoil of the heart muscles lowers pressure in the atria and ventricles. Blood is drawn into the heart from the arteries and veins, closing the semilunar valves.

Quick Questions

Q1 Name the chamber of the heart:
  a which pumps blood from the heart to the lungs
  b which receives blood from the vena cavae
  c which pumps blood from the heart to the aorta.
Q2 Give *three* reasons why *Daphnia* is a suitable subject for study in the heart rate and caffeine experiment.

Thinking Task

Q1 Draw a flowchart for the cycle of events on the right side of the heart during one 'turn' of the cardiac cycle. Use the flowchart of the events on the left side and the written account of what happens on the right side to help you with this.

Cardiovascular disease (CVD)

Blood clotting is a vital defence mechanism for the body. If you suffer a cut or graze, then clotting can:

- minimise blood loss
- help prevent the entry of pathogens
- provide a framework for repair.

But, if a clot occurs inside a blood vessel it can be very dangerous, blocking blood flow and sometimes leading to the death of tissues.

Clot formation is stimulated when there is damage to a blood vessel. Damage exposes collagen fibres to which **platelets** (small cells with no nucleus formed when a precursor cell fragments) attach. The platelets release a clotting factor called **thromboplastin**. In the presence of calcium ions and vitamin K, thromboplastin converts inactive prothrombin into active thrombin. This in turn converts the soluble fibrinogen into insoluble fibrin, which forms a network of fibres, trapping cells and debris to make a clot.

ResultsPlus
Watch out!

Make sure you know that the two substances, prothrombin and fibrinogen, are the inactive forms. *Pro* means, roughly, first as in *pro*totype. You may have met trypsinogen at GCSE and know that the *–ogen* suffix means the inactive form.

Atherosclerosis

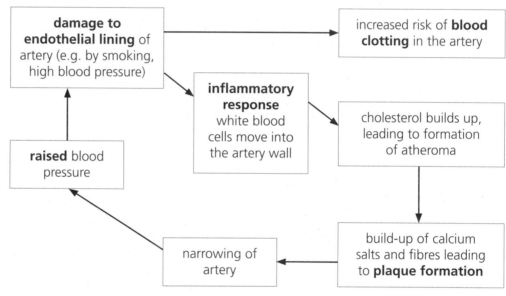

The formation of an atherosclerotic plaque is a positive feedback phenomenon.

ResultsPlus
Examiner tip

Athero (artery) sclerosis (hardening) – the plaque makes the wall become less elastic and narrows the artery.

CVD

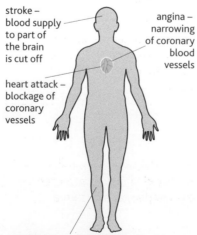

stroke – blood supply to part of the brain is cut off

angina – narrowing of coronary blood vessels

heart attack – blockage of coronary vessels

peripheral vascular disease – thrombosis, narrowing of arteries to periphery, especially common in the legs

The risk of suffering from CVD is increased by:

- genetic factors
- age
- gender
- high blood pressure
- lifestyle factors – diet, exercise and smoking.

(See page 16 for more details.)

The various types of cardiovascular disease (CVD), all caused as a result of atherosclerosis.

Treatment of CVD

Risk of CVD can be reduced by lifestyle changes:

- stopping smoking
- moderate exercise several times a week
- stopping over-consumption of alcohol
- dietary changes, especially lowering cholesterol and saturated fat intake.

Medical treatments which can help are:

- reducing high blood pressure (usually defined as over 160 (systolic)/100 (diastolic)) by **antihypertensives**
- reduction of blood cholesterol, e.g. by diet or by drugs such as **statins**
- **anticoagulants**
- **platelet inhibitors**.

The table summarises some drugs used for treating CVD and the risks associated with their use.

Drug treatment	Mode of action	Risks/side effects
diuretics (antihypertensive)	increase volume of urine; lowers blood volume and pressure	very occasional dizziness, nausea, muscle cramps
beta blockers (antihypertensive)	block response of heart to hormones and make contractions less frequent and less powerful	possible link with diabetes
ACE inhibitors (antihypertensive)	block the production of angiotensin (ACE stands for angiotensin converting enzyme) which normally causes arterial constriction and a rise in blood pressure	cough, dizziness, heart arrhythmia, impaired kidney function
statins	lower cholesterol level in the blood by blocking the liver enzyme that makes cholesterol	muscle aches, nausea, constipation and diarrhoea; very rarely inflammation reactions can occur which, even more rarely, are fatal; also, again rarely, liver failure; also, people may stop trying to eat a healthy diet, leaving it all to the statins
anticoagulants, e.g. warfarin	reduce risk of clot formation	risk of uncontrolled bleeding; dosage control is essential
platelet inhibitory drugs, e.g. aspirin, clopidogrel	make platelets less sticky	aspirin irritates the stomach lining and can cause serious stomach bleeding; using clopidogrel with aspirin can make the risk even greater

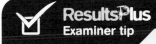

ResultsPlus
Examiner tip

You probably will not know enough about what these treatments do to work out what the risks might be, so try to learn them thoroughly.

Quick Questions

Q1 List the *factors* which must be present in order for a blood clot to form.

Q2 Even if all the blood clotting chemicals you have listed in Q1 were present, this would not allow a clot to form. What else is needed, apart from a range of chemicals?

Thinking Task

Q1 Use the outline flowchart below to summarise the clotting process by substituting the correct terms from the description on page 12.

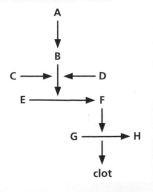

Structure and function of carbohydrates

Carbohydrates contain carbon, hydrogen and oxygen in the ratio $(CH_2O)_n$. They are made of saccharide units (literally sugar) – a prefix (mono-, di-, poly-) denotes how many units.

Sugar	Basic structure	Examples	Notes
monosaccharides (single saccharides)	mono-	glucose	• main substrate for respiration • soluble, osmotic effect
		galactose	• soluble, osmotic effect
		fructose	• 'fruit sugar' • soluble, osmotic effect
disaccharides (two saccharides)	di-	sucrose	• glucose + fructose • main transport sugar in plants • soluble
		lactose	• 'milk' sugar • glucose + galactose • soluble
		maltose	• glucose + glucose • soluble
polysaccharides (many saccharides)	poly- unbranched	amylose	• found in starch – energy storage molecule in plants • α-glucose* molecules in tight spirals so compact • insoluble, no osmotic effect
	poly- branched	amylopectin	• found in starch – energy storage molecule in plants • branched chains of α-glucose molecules – lots of terminal ends so digested more rapidly than amylose • insoluble, no osmotic effect
	poly- branched	glycogen	• energy storage molecule in animals, bacteria and fungi • branched chains of glucose molecules • compact, insoluble, no osmotic effect

α- and β-glucose forms are covered on page 80.

α- and β-glucose forms are covered on page 80.

The bonds that form between monosaccharides are **glycosidic bonds**. In the reaction that forms a glycosidic bond there is a loss of one molecule of water (called a **condensation reaction**).

Hydrolysis breaks the glycosidic bond, adding back a molecule of water. This is catalysed by enzymes.

ResultsPlus
Watch out!

Do not use the word *sugar* without qualifying it. In everyday English, sugar is the disaccharide *sucrose*, but scientifically there are many different sugars. Make it clear what you mean.

ResultsPlus
Examiner tip

Saccharides, whether di- or poly-, are always joined together by glycosidic bonds.

Hydrolysis of maltose by addition of water and condensation of two glucose units to form maltose with loss of water.

Thinking Task

Q1 Relate structure to function in glucose, sucrose, amylose and amylopectin.

Quick Questions

Q1 Lactose, sucrose and maltose are all disaccharides, how do they differ from each other?

Structure and function of lipids

Lipids are biological molecules that are insoluble in water but soluble in organic solvents like ethanol. Triglycerides are lipids made of three fatty acids and one glycerol molecule joined via an ester bond.

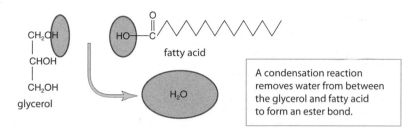

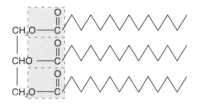

three ester bonds in a triglyceride
formed from glycerol and three fatty acids

Three ester bonds form between the glycerol and three fatty acid molecules. Each bond is formed in a condensation reaction in which a molecule of water is lost. The reaction is catalysed by an enzyme.

Variation in triglycerides

All glycerol molecules are the same. The fatty acids vary in:
- the length of the hydrocarbon chain, from eight or so to 20 carbons
- the absence or presence, and number, of double bonds
- the mix of fatty acids (i.e. all three the same, all different or some other combination).

Two carbons can be joined by a double bond, but this means that fewer hydrogens can bind and thus the molecule is less saturated with hydrogen.

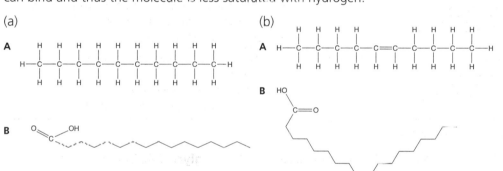

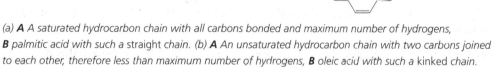

*(a) **A** A saturated hydrocarbon chain with all carbons bonded and maximum number of hydrogens, **B** palmitic acid with such a straight chain. (b) **A** An unsaturated hydrocarbon chain with two carbons joined to each other, therefore less than maximum number of hydrogens, **B** oleic acid with such a kinked chain.*

The difference in saturation has important effects on triglycerides:
- structure   • physical properties   • biological significance.

Saturated (fats)	Monounsaturated (one double bond) (oils)	Polyunsaturated (more than one double bond) (oils)
strong intermolecular bonds	intermolecular bonds weaker because of kinked shape	intermolecular bonds weaker because of kinked shape
solid at room temperature	liquid at room temperature	liquid at room temperature
straight chain molecule	molecule with one kink in chain	molecule with x kinks in chain (x = no. of double bonds)

Quick Questions

Q1 What is the similarity between the formation of a glycosidic bond between monosaccharides and that of an ester bond between glycerol and fatty acids?

Q2 When a triglyceride is broken apart by hydrolysis, how many components are found as products? Name them.

Thinking Task

Q1 How many water molecules would it take to fully hydrolyse one triglyceride to its component fatty acids and glycerol? Explain your answer.

Risk factors for cardiovascular disease

Risk is *the probability of the occurrence of an unwanted event or outcome.* When formally assessing a risk, it is usual to multiply the likelihood by the severity to give a **risk rating**.

We often under- or overestimate risks. A **perceived risk** is based more on factors such as interest and approval of the activity than on anything mathematical.

The risk of developing a health condition is determined from many studies, but people often ignore the advice which is implied by these studies. For CVD, the factors which increase the **actual risk** of suffering are:

Factor	Examples
genetic	Tendency to high blood pressure and poor cholesterol metabolism. Arteries that are more easily damaged. Mutations in genes that affect relative HDL:LDL levels in blood.
gender	Oestrogen gives women some protection from CVDs before the menopause. Then the risk in both sexes is about the same.
ageing	Elasticity and width of arteries decrease with age.
diet	Many **correlations** (links) between dietary habits and level of CVD, e.g. saturated fat, cholesterol and lipoprotein levels. Some evidence that these correlations are **causal**, particularly for blood cholesterol levels.
high blood pressure	Very important – should not be sustained >140 mm Hg systolic and 90 mm Hg diastolic (expressed as 140/90).
smoking	Correlation *and* causation shown because chemicals in smoke physically damage artery linings and also cause them to constrict.
inactivity	Regular vigorous exercise reduces the risk of CVD by reducing blood pressure and raising HDL (good cholesterol) levels.
obesity	Increases risk of CVD and developing type II diabetes.

ResultsPlus
Examiner tip ✓

Correlation does *not* imply causation. This is one of the most important facts you need to remember as a scientist (see page 18).

Why do we think high fat intake *causes* CVDs?

Studies have shown a strong correlation between high intake of saturated fats and high blood cholesterol. Cholesterol helps form plaques; this is the *causal* link.

The cholesterol story is not simple. Cholesterol is a water-insoluble lipid. For transport, it has to be carried by proteins in soluble complexes called lipoproteins. There are two kinds of lipoprotein:

Low density lipoprotein (LDL, 'bad' cholesterol)	High density lipoprotein (HDL, 'good' cholesterol)
formed from saturated fats, protein, cholesterol	formed from unsaturated fats, protein, cholesterol
bind to cell surface receptors, which can become saturated leaving the LDLs in the blood	transport cholesterol from body tissues to liver where it is broken down.
associated with formation of atherosclerosis	reduces blood cholesterol levels, discourages atherosclerosis
should be maintained at low level	should be maintained at high level

ResultsPlus
Watch out! !

'Good' and 'bad' cholesterol are no different to each other chemically, it is the way the cholesterol is carried in the blood that differs.

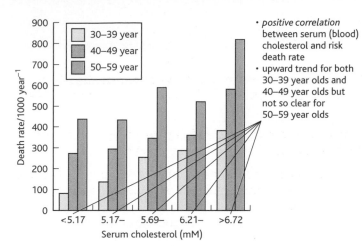

- *positive correlation* between serum (blood) cholesterol and risk death rate
- upward trend for both 30–39 year olds and 40–49 year olds but not so clear for 50–59 year olds

Effect of total blood cholesterol on deaths from coronary heart disease (Framingham Study).

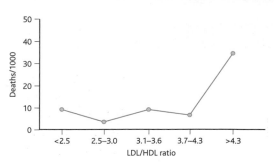

Since LDL and HDL cholesterol have opposite effects, the ratio of LDL to HDL appears to be a very important indicator of risk from CVD.

Energy balance

The energy you need in a day depends on your Basal Metabolic Rate (BMR) and level of activity. BMR varies with gender, age and body mass.

Worked Example

Fred, a carpenter, weighs 70 kg. The human basic energy requirement is $4 \, kJ \, kg^{-1}$ body mass h^{-1}.

So Fred has a *basic* energy requirement of: $4 \times 70 = 280 \, kJ$ per hour

His BMR is: $280 \times 24 = 6720 \, kJ \, day^{-1}$

Daily activity for an average carpenter uses 5000 kJ

Fred's total energy need per day is: $6720 + 5000 = 11720 \, kJ$

Fred's daily energy balance is: $10300 - 11720 = -1420 \, kJ$

He is using more than he is consuming and will lose weight.

Bob, another carpenter, consumes $13700 \, kJ \, day^{-1}$. Bob's daily balance is: $13700 - 11720 = 1880 \, kJ$

So, he will put on weight and may become overweight or obese.

Fred's daily energy intake

1800 kJ breakfast

2500 kJ lunch

4000 kJ supper

2000 kJ in drinks and snacks

$= 10300 \, kJ \, day^{-1}$ total

Around 50% of the UK adult population is overweight, 20% are obese. Body Mass Index (BMI) is one obesity indicator used to define these conditions. It is given by:

$$BMI = \frac{body \ mass \ in \ kg}{(height \ in \ m)^2}$$

For example, if weight = 70 kg and height = 1.75 m

$$BMI = \frac{70}{1.75^2} = \frac{70}{3.06} = 22.9$$

BMI	Status
<20	underweight
20–25	correct weight
25.1–30	overweight
>30	obese

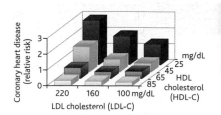

Population studies on risk factors

Epidemiologists are scientists who carry out research to try to determine the risk factors for health.

There are two kinds of epidemiological study.

Cohort studies
Main features:
* large number of people followed
* long period of time
* monitored to see if they develop the condition under study
* those who develop the condition are put in one group and those who do not in another
* various risk factors that subjects have been exposed to are looked at (by interviewing)
* correlations searched for.

For example, the ongoing Framingham study has involved thousands of participants from 1948 onwards. This study shows that age, total cholesterol and HDL, smoking and high blood pressure are the major risk factors for CVD.

Case-control studies
Main features:
* A group with the condition (cases) is compared with a group who do not have it (control).
* Past history is then investigated to try to identify factors leading to one group having the disease and the other not.
* It is very important in this case to match the control group with the case group for such things as age and gender.
* This is rather like the classic experimental design where independent variables are controlled.

There is no famous example from CVD research, but in the studies by Doll and Hill on lung cancer, a group with lung cancer was compared with a group without. A correlation with smoking was found.

ResultsPlus
Watch out!

Correlation does *not* imply causation.

Lung cancer		No lung cancer	
Smokers	**Non-smokers**	**Smokers**	**Non-smokers**
647	2	622	25

Cigarette smoking history of those with and without lung cancer in the study by Doll and Hill.

Remember, you can have correlation without causation (often!) but never the reverse. The fact that two variables are correlated suggests there *might* be a causal link, not that there is one. To show causality, a mechanism by which one factor causes the other needs to be proved. This may require an experiment.

A study showing a correlation between blood cholesterol and incidence of heart attacks in the UK gives a graph like this:

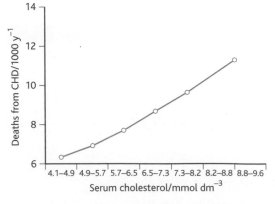

Relationship between blood cholesterol levels and death from CHD (coronary heart disease, a type of CVD).

However, this does not justify us in saying that high blood cholesterol causes CHD!

What makes a 'good' study?

The following are key features of a study which collects **valid** and **reliable** data:

- The studied sample should be representative of the whole population, otherwise the sample could be **biased**.
- Variables should be controlled as far as possible when selecting cohorts or control groups. This is one of the most difficult aspects of this kind of study since human beings are so variable in terms of genes and environment. Measurement techniques or the questions on a questionnaire must be standardised.
- Sample size is very important. In many diseases, only a low percentage of the population have the condition so an apparently large sample size might only contain a small number of individuals with the condition.

ResultsPlus
Watch out!

Valid data are data that accurately measure what they are supposed to, and do so every time, making the data reliable.

Quick Questions

Q1 Compare cohort studies with case-control studies.
Q2 List the key features of a good study used to determine health risk factors.

ResultsPlus
Examiner tip

Although it is not generally possible to experiment on people, it is possible to select a sample carefully so as to ensure that non-experimental variables, such as age and sex, are matched so it is more like a traditional controlled experiment in the laboratory.

Thinking Task

Q1 Look at these data about the rate of CHD (coronary heart disease, an aspect of CVD) in France and the UK and some information about some of the key factors known to be involved in CHD from other research.

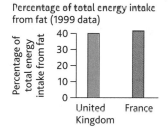

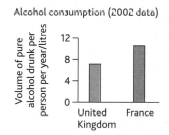

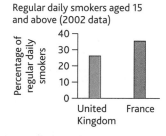

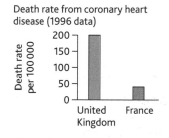

a Why would you expect CHD rates in France to be higher than those in the UK?
b Suggest why the French CHD rates are not what we would expect.

Using scientific knowledge to reduce risk

A lot is now known about the effect of diet, exercise and smoking on the risk of CVD, but many people do not change their lifestyle due to **perception of risk**, which is affected by:

- own experience, which will carry more weight than statistics: 'my auntie smoked 20 a day until she was 90, and died falling downstairs; not from lung cancer!'
- inability to assess risks well, for example many believe that smoking keeps weight down, but the relative risks of obesity and smoking are not considered
- peer pressure – very powerful, for example in the case of alcohol consumption and smoking among young people
- the idea that if something is destined to happen then it will and there is not much one can do about it
- the remoteness of the likely consequence – a bag of chips now can give much pleasure and it is easy to trade that off against the increased risk of a heart attack in 30 years!

However, some progress is being made. The huge increase in the popularity of low cholesterol/low fat spreads, the fall in incidence of smoking and the rise in the popularity of gyms and other health clubs all attest to this.

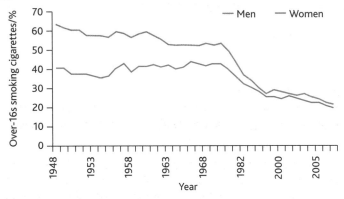

Incidence of smoking in over-16s in the UK, 1948–2007.

Vitamin C

Antioxidants, found in fresh fruit and vegetables, are important in neutralising free radicals which can damage cells including those in the cardiovascular system. One well-known antioxidant is ascorbic acid (vitamin C). Heat treatment may destroy some or all of the vitamin C, so cooking can reduce the health value of a food. Vitamin C content is easy to find out with a simple practical procedure based on the following:

- Antioxidants are reducing agents as they lose electrons.
- Many substances will change colour when they are reduced; these substances are called REDOX dyes (reduction–oxidation dyes).
- DCPIP is a blue dye in its non-reduced form; it goes colourless when it gains electrons.

Worked Example

1 Calibrate the DCPIP – add a measured volume of a known concentration of vitamin C solution to the DCPIP.
- Assume it took $0.6\,cm^3$ of a 1% vitamin C solution to decolourise the DCPIP.
- In a 1% vitamin C solution, there is $1\,g\,100\,cm^{-3}$ so there are $10\,mg$ of vitamin C in $1\,cm^3$.
- In $0.6\,cm^3$ of the solution, there are $6\,mg$ of vitamin C.

Worked Example

2 Add the unknown solution drop by drop to the DCPIP until it has gone colourless.

Juice tested	Average volume of juice required to decolourise DCPIP /cm^3
grapefruit juice	1.61
pineapple juice	11.56
orange juice	2.12
orange drink	1.45
fresh lemon juice	1.73
bottled lemon juice	24.0

The vitamin C content of the grapefruit juice is worked out like this:

6 mg vitamin C decolourised 1 cm^3 of DCPIP

1.61 cm^3 of the grapefruit juice decolourised 1 cm^3 of DCPIP so 1.61 cm^3 of grapefruit juice contains 6 mg vitamin C

Therefore 1 cm^3 of grapefruit juice contains:

$$\frac{6}{1.61} = 3.7 \text{ mg of vitamin C.}$$

All experiments are subject to errors:
- **systematic errors** – values differing from the true value by the same amount
- **random errors** – values lying randomly above or below a true value.

Possible systematic errors in this experiment:
- The burette or pipette from which the vitamin C is delivered has an error in its volume. (For example, glass expands and contracts with a rise and fall in temperature. A pipette or burette is designed to be accurate at a particular temperature. If the room temperature is above or below this, a systematic error will result.)
- The experimenter reads the volume at the wrong place each time, so either under- or overestimates the true volume added each time.

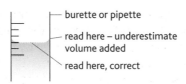

- burette or pipette
- read here – underestimate volume added
- read here, correct

Possible random errors in this experiment:
- The end point is misjudged, as it is quite tricky to say exactly when the DCPIP has become colourless.
- The vitamin C cannot be added by less than a drop a time, so sometimes the next drop may be too much, other times too little.

? Quick Questions

Q1 List *three* factors which affect perception of risk.

Q2 Copy and complete this passage:
Vitamin C is an _____ which gives electrons to other substances. This make its an _____ agent. DCPIP is a _____ agent which can gain _____ from _____ agents.

 **Thinking Task**

Q1 Calculate the vitamin C content of the juices in the table in the worked example above.

Topic 1 – Lifestyle, health and risk checklist

By the end of this topic you should be able to:

Revision spread	Checkpoints	Spec. point	Revised	Practice exam questions
Transport and circulation	Recall why water is important in transport.	LO2	☐	☐
	Explain why animals have a heart and circulation to overcome limitations of diffusion.	LO6	☐	☐
	Explain the adaptations of blood vessels to their function.	LO8	☐	☐
Cardiac cycle and heart rate	Describe the cardiac cycle and relate structure to function in the heart and its associated blood vessels.	LO7	☐	☐
	Describe a practical investigation on the effect of caffeine on heart rate in *Daphnia* (core practical).	LO9, LO1	☐	☐
Cardiovascular disease	Describe the blood-clotting process and explain how it might be involved in CVD.	LO10	☐	☐
	Explain the events which lead to atherosclerosis.	LO11	☐	☐
	Describe the benefits and risk of treatments of CVD treatments.	LO13	☐	☐
Structure and function of carbohydrates	Describe and compare the various types of carbohydrates and relate their structure to their function.	LO3	☐	☐
	Describe how carbohydrates form polymers and how these can be broken down.	LO4	☐	☐
Structure and function of lipids	Describe the formation of triglycerides, and compare saturated and unsaturated lipids.	LO5	☐	☐
Risk factors for cardiovascular disease	Describe the risk factors for CVD.	LO12	☐	☐
	Analyse and interpret data on the possible link between cholesterol, HDL and LDL levels and health, and the evidence for a causal relationship between total cholesterol, LDL and CVD.	LO14	☐	☐
	Analyse data on human energy budgets and diet, and discuss the consequences of energy in balance.	LO17	☐	☐
	Explain why risk perception is often different from actual risk.	LO20	☐	☐
Population studies on risk factors	Analyse data on mortality rates to determine health risks and understand the difference between correlation and causation.	LO18	☐	☐
	Evaluate the design of studies to determine health risk factors.	LO19	☐	☐
Using scientific knowledge to reduce risk	Discuss how people use scientific knowledge about the effect of diet, exercise and smoking to reduce risk of CVD.	LO15	☐	☐
	Describe how to investigate the vitamin C content of food and drink (core practical).	LO16	☐	☐

ResultsPlus
Build Better Answers

1 Describe the structure of starch and explain why this structure makes it a suitable molecule for storing energy. (4)

✓ **Examiner tip**

Make sure you read and take note of *all* parts of a question. This one requires you to discuss structure *and* relate it to function, a common requirement.

Student answer	Examiner comments
There are two types of starch, amylose which forms 20–30% of starch and amylopectin which makes up 70–80%. Amylose has a helical structure which means that it is compact. However there are no end branches, meaning that it is not able to be used for a fast supply of energy – it has a long structure of between 2000 and 5000 units, meaning it stores a lot of energy. Amylopectin has more end branches allowing for rapid energy use and is also very compact, with 1,4 and 1,6 bonds every 20–30 units.	This is quite a good answer, with a number of the structural points mentioned. However, there is limited attempt to relate structure to function. This question carried 4 marks, but a maximum of 3 of these were awarded for purely structural points. ■ A **basic answer** would make just one point, such as referring to the existence of the two forms amylose and amylopectin, and make no reference to function. ▲ An **excellent answer** would make three good structural points, the amylose/amylopectin one and maybe describe each of these (helical and branched respectively). It would then make a functional point such as that starch is insoluble and does not easily move out of cells or have an osmotic effect due to this.

2 A study was carried out into the number of cigarettes smoked by men per year and the number of deaths from lung cancer. The graph shows the results of this study.
Describe the changes in the number of deaths from lung cancer between 1920 and 1975. (3)

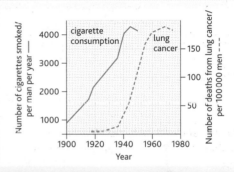

✓ **Examiner tip**

In this kind of question you should be looking to identify trends or changes in trends and you should always aim to manipulate some data to back up your description.

Student answer	Examiner comments
At the start there was no change in the number of deaths from lung cancer. There was a slight increase in number of deaths after the mid 20s which got steeper. In 1969 the line dropped.	This quite a basic answer and would achieve only 1 mark. The first sentence makes a good point but it is not related to any dates so gains no mark. The second sentence gains the 1 mark this answer achieves by implying, although not very clearly, that the trend was overall upward. The reference to the mid 20s is too imprecise for a mark which required a date range for the slight increase (1924–1936/7). The last point made is mathematically correct, but tells us nothing about the deaths, just a line. ■ A **basic answer** would note the overall increase and nothing more. ▲ An **excellent answer** would describe the finer points within the graph, no change from 1920–1924, a gradual increase from 1924–1936/7, sharp increase between 1936/7 and 1955, 1960 or 1969 and the sharp fall after 1969. That is, dates would be quoted in every case. A further way of gaining a mark would be to do some manipulation on the figures, for example to say that the number of deaths increased from 5 deaths per 100 000 men in 1920 to 190 deaths per 100 000 men in 1969, an increase of 185 deaths per 100 000 men.

Practice questions

1 Carbohydrates are compounds that include monosaccharides, disaccharides and polysaccharides.

(a) (i) The table below lists some features of four carbohydrates.
Copy and complete the table below by ticking the relevant boxes. The first line has been done for you.

Feature	Glucose	Glycogen	Maltose	Starch
1–4 glycosidic bonds present	✓	✓	✓	✓
1–6 glycosidic bonds present		✓		✓
made up of many monomers		✓		✓

(4)

(ii) Name the disaccharide made up of α-glucose and galactose. Lactose (1)

(iii) The diagram below shows a disaccharide molecule.

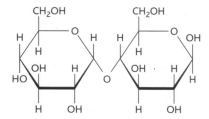

Draw the molecules resulting when this disaccharide molecule is split into its two component monosaccharides. (2)

(iv) Name this type of reaction. (1)

(b) Explain the advantages of glycogen as an energy storage molecule in the human body. (5)

(Sample Assessment Materials for 6BI01, Q1)

2 The graph below shows death rates from coronary heart disease (CHD) in men from 1970 to 1995 in four countries.

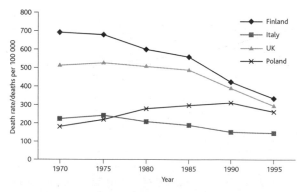

(a) Describe the changes in death rates shown on the graph. (3)

(b) The graph below shows the percentage of men from these countries in 1980 with high systolic blood pressure.

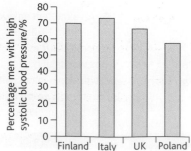

It has been suggested that there is a link between high blood pressure and deaths from CHD.

(i) Using both graphs, give *two* pieces of evidence to support this hypothesis. (2)

(ii) Suggest how the data shown in the graphs do not fully support this hypothesis. (2)

(c) Suggest how high blood pressure can result in less oxygen reaching heart muscle. (3)

(Sample Assessment Materials for 6BI01, Q2)

3 (a) If a person routinely takes in more energy from their food than they use up, they will have a positive energy balance. This results in the person becoming overweight or even obese.

$$\text{body mass index (BMI)} = \frac{\text{body mass}}{\text{height}^2}$$

BMI (kg m^{-2})	Classification of body weight
less than 20	underweight
20–24.9	normal
25–29.9	overweight
30–40	obese
greater than 40	severely obese

(i) Calculate the body mass index for a person with a body mass of 95 kg and a height of 1.70 m. Show your working. (2)

(ii) Suggest and explain the advice a doctor might provide to a person with a BMI value of over 40. (5)

(b) An individual's estimated average requirement for energy (EAR) can be determined by multiplying the basal metabolic rate (BMR) by the physical activity level (PAL).

EAR = BMR × PAL

The graph shows the effect of age on EAR values for males with low levels of physical activity.

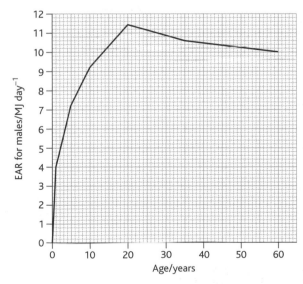

(i) Use the graph to estimate the EAR of a 35-year-old man with a low level of physical activity. (1)

(ii) The PAL value for a man with a low level of physical activity is 1.4. Calculate the BMR for a 35-year-old man with this level of physical activity. Show your working. (2)

(SNAB AS, Jan 2005, Q 7)

Structure of amino acids and proteins

Proteins are polymers made of many similar molecules (amino acids) joined together in long chains. These chains are then folded to give the protein a 3-D shape.

What is an amino acid?

An **amino acid** consists of a central (or alpha) carbon atom attached to an amino group, NH_2, a carboxylic acid group, COOH, a hydrogen atom and a variable side group (called the R group). The R can represent one of 20 different side chains.

General structure of an amino acid.

In naturally occurring proteins there are 20 different amino acids (i.e. 20 different R groups). The simplest R group is hydrogen in the amino acid glycine and one of the most complex is in tryptophan.

Proteins have *four* levels of structure.

1 Primary structure

Amino acids are able to join to other amino acids to form chains:
- **dipeptide** – two amino acids bonded together
- **polypeptide** – chain of many amino acids.

These chains always have an amino group at one end and a carboxylic acid group at the other, and so can join with more amino acids at either end to form chains of theoretically infinite length.

The bond between amino acids is a **peptide bond** and is formed by the loss of water in an enzyme-catalysed **condensation reaction**.

Formation of a peptide bond between two amino acids.

The sequence of amino acids in the polypeptide chain is the primary structure of the protein.

2 Secondary structure

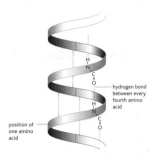

The amino and carboxylic acid groups in the chain carry small amounts of charge, negative on the CO of the carboxylic group and positive on the NH of the amino group. These charges can result in hydrogen bonds forming.

Many hydrogen bonds can hold together:

- a helical structure (the α helix, see left)

- sheet made up of polypeptides laid out parallel to each other (β pleated sheet, see right).

3 Tertiary structure

Further folding forms a precise three-dimensional shape held together by bonds between R side chains as well as hydrophobic interactions. These are:

- hydrogen bonds
- ionic bonds between ionised R groups
- covalent bonds, such as those between sulfur groups in cysteine called a disulfide bridge
- polar interactions – groups that are water-loving (hydrophilic or polar) arrange themselves on the outside of the protein and those that are hydrophobic (water hating or non-polar) are on the inside.

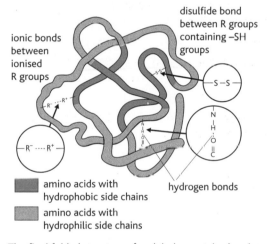

The final folded structure of a globular protein showing all the bonds and interactions which hold it together.

4 Quaternary structure

Two or more polypeptide chains held together by hydrogen bonds, as in haemoglobin.

Function related to structure

Proteins have many functions and work because of their shape:

- Globular proteins have a complex tertiary, and sometimes quaternary, structure, e.g. enzymes, membrane proteins, antibiotics, oxygen transport pigments.
- Fibrous proteins have little or no tertiary structure and parallel polypeptide chains are cross-linked to form fibres, e.g. keratin in skin and nails, collagen in skin.

Quick Questions

Q1 Name *three* different types of pairs of molecules that can be joined together in a condensation reaction.

Q2 Which type of bond involved in forming proteins will remain intact when it is fully denatured?

> ### ResultsPlus
> **Watch out!**
>
> Proteins can be:
> - broken down (or **hydrolysed**) when the peptide bonds between amino acids are broken. This is what happens when a protein is digested in the gut
> - **denatured** when the 3-dimensional shape is disrupted. This can be due to breakage of H bonds when the molecule vibrates more at higher temperatures, or changes in pH which break ionic bonds between R side chains.
>
> Do not mix these two up with each other.

Thinking Task

Q1 Are there any *two* amino acids that cannot join together? Explain your answer.

Enzyme action and rates of reaction

Catalysts are substances that speed up reactions. Without catalysts, every chemical reaction in a living thing would proceed far too slowly.

Enzymes are biological catalysts that speed up reactions both inside cells (intracellular) and outside cells (extracellular). Intracellular reactions include all those involved in protein synthesis, such as peptide bond formation. Extracellular reactions include digestion, and decomposition by bacteria.

Enzymes are globular proteins – each enzyme has a specific 3D shape including an **active site**. Only molecules with a specific shape (the **substrate**) fit into the active site. The close interaction of the complementary-shaped enzyme and substrates is known as the **lock and key hypothesis**. The substrate may induce the enzyme into the right shape – this is called the **induced fit theory**.

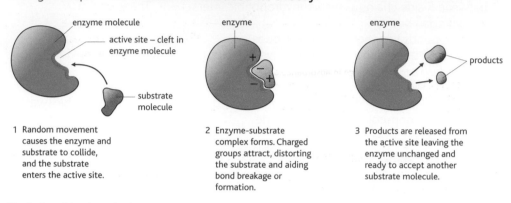

1 Random movement causes the enzyme and substrate to collide, and the substrate enters the active site.

2 Enzyme-substrate complex forms. Charged groups attract, distorting the substrate and aiding bond breakage or formation.

3 Products are released from the active site leaving the enzyme unchanged and ready to accept another substrate molecule.

The lock and key hypothesis.

To start a reaction, bonds need to be broken. This can be done by:
- adding heat energy
- adding a catalyst.

Molecules in a mixture have differing amounts of energy. Only those with enough energy will react.

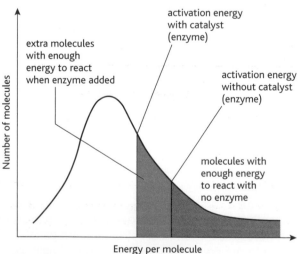

*The concept of **activation energy**.*

Enzymes lower the activation energy by:
- assisting the breaking or making of bonds as charged groups in the active site interact and distort the shape of the substrate(s).
- creating a pH within the active site which makes the reaction more likely.

Rates of reaction

Enzyme-catalysed reactions are very fast. The rate of reaction slows down as substrate is used up. To measure the initial rate of reaction, graph the results and measure the slope of the line before it starts to become non-linear.

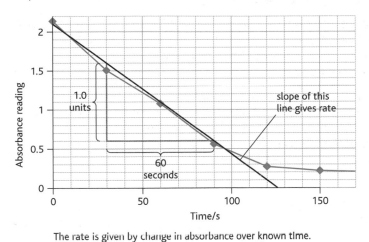

In this experiment, a protease was used to break down a cloudy protein solution. The decrease in cloudiness was measured with a colorimeter, taking an absorbance reading every few seconds.

The rate is given by change in absorbance over known time.
Change in absorbance is = 1.0 absorbance units
Time over which this happens = 60 s
Rate = 1.0/60 = 0.017 absorbance units s⁻¹ = 1.02 absorbance units s⁻¹

Measuring initial rate of reaction.

Repeating the reaction with other enzyme concentrations gives initial rates for each concentration. Plotting these results gives a graph showing the effect of enzyme concentration on initial reaction rate.

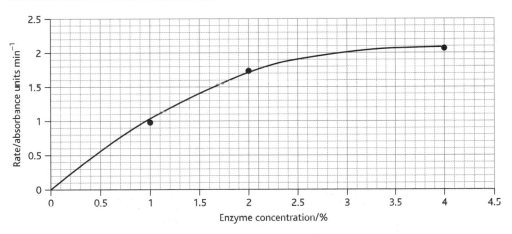

'Rate' vs. enzyme concentration for reaction of protease described above.

Worked Example

Describe and explain the curve of enzyme concentration on initial reaction rate.

Describe: As the concentration of enzyme increases, so does the initial rate of the reaction, but a point is reached when this stops happening and the curve levels off.

Explain: This is because, when there is not very much enzyme and lots of substrate, every active site is occupied by substrate. Add more enzyme and there are more active sites and still plenty of substrate, so the rate goes up. Eventually adding more enzyme achieves nothing as there is now not enough substrate to fill all the active sites all the time.

Cell membranes

All cells are surrounded by a partially permeable **membrane** that controls the movement of substances into and out of the cell, helps to 'glue' one cell to another and acts as a receptor surface for such things as hormones. Simple chemical tests show the presence of protein, lipid (phospholipid), cholesterol and carbohydrate in membranes.

Building a model of membrane structure

Any model of membrane structure must take into account known facts about membranes and their components.

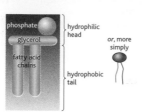

A phospholipid molecule has a hydrophilic (water-loving) head (glycerol and a phosphate group) and two hydrophobic (water-hating) tails.

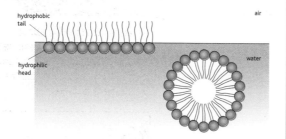

(a) Monolayer and spherical micelle; (b) bilayer membrane, two molecules thick with hydrophobic tails inside the bilayer. The protein molecules that form channels through the membrane are lined with hydrophilic regions of the protein molecule, with hydrophobic regions facing the lipid bilayer.

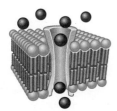

A protein channel.

Fact to explain	Implications for the model
Phospholipids are like triglyceride lipids, they are **polar** with a hydrophilic head and hydrophobic tail.	• have two different responses to water
When suspended in water, phospholipids naturally form bilayers with hydrophobic tails inside and hydrophilic heads outside.	• suggests phospholipid bilayer structure
Experiments on the total area of a monolayer film of phospholipids extracted from cells compared to the surface area of the cells showed that the film was twice as large as the cell surface area.	• further supports bilayer model
In electron microscope images of cell surfaces, proteins can be seen sticking out.	• proteins not in continuous sheet on membrane surfaces, but form a **mosaic** amongst the lipids
When lectins, which only react with carbohydrates, are added to a membrane they are found only on the outside.	• carbohydrates found only on outside of membrane
Some water-soluble substances pass into and out of cells.	• suggests proteins in membrane act as channels for movement of substances in and out
Ionic and polar molecules do not pass easily through membranes, but lipid-soluble substances do.	• membranes are made mainly of lipid

The accepted model of cell membrane structure is the fluid mosaic model, which explains all the facts in the table.

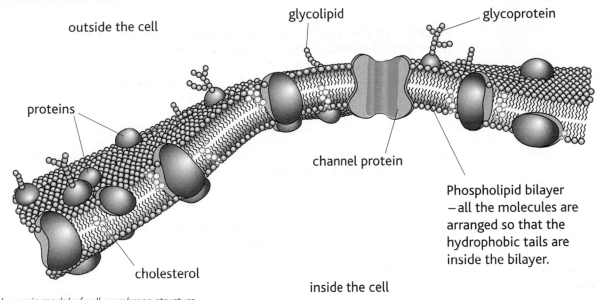

The fluid-mosaic model of cell membrane structure.

Quick Questions

Q1 What are the chemical components of a membrane?

Q2 State how many subunits a phospholipid would break down into on hydrolysis and name them.

ResultsPlus
Watch out!

When asked to write about membranes, many students forget that they are found in structures within the cell and not just the cell surface membrane.

Thinking Task

Q1 As you work through your revision make a list, a table or a mind map (spider diagram) of functions of membrane proteins other than as channels.

Transport across cell membranes

Molecules move through cell membranes by various methods depending on the properties of the molecule and the needs of the cell.

Types of movement

Diffusion:

- small and/or non-polar lipid-soluble molecules, e.g. oxygen, carbon dioxide
- directly through phospholipid bilayer
- passive
- net movement down concentration gradient.

Facilitated diffusion:

- As diffusion but requiring a **channel protein** in the membrane to allow (facilitate) the diffusion of polar molecules, charged and water-soluble groups.

Osmosis:

- essentially the diffusion of free water molecules
- net movement of free water molecules from a solution with lower solute concentration to a solution with a higher solute concentration
- movement through a partially permeable membrane.

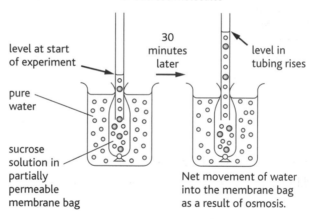

○ Water molecules
◉ Sucrose molecules

level at start of experiment

30 minutes later

level in tubing rises

pure water

sucrose solution in partially permeable membrane bag

Net movement of water into the membrane bag as a result of osmosis.

A simple cell model in which the water moves into the bag at a greater rate than it moves out, so the water rises up the tube. Sucrose molecules cannot move through the walls of the bag because they are too big to pass through the partially permeable material, just like the cell membrane.

Active transport:

- all kinds of molecules possible
- through **carrier proteins**
- active – needs energy from the breakdown of ATP
- *up* a concentration gradient.

Exo- (out of cell) and endocytosis (into cell):

- large particles of all kinds
- bulk transport involving vesicles made from cell surface membrane in endocytosis, and fusing with the cell surface membrane in exocytosis.

Investigating membranes

The fluid mosaic model of membrane structure suggests that temperature might affect cell membranes. As temperature increases, phospholipids will become more fluid, allowing molecules to leak from the cell. Plants containing pigment within the cytoplasm and/or vacuole can be used to test this idea, e.g. beetroot.

Cut equal sized pieces of tissue.

↓

Rinse under running water to remove all betalain (pigment) released by cutting.

↓

Place the pieces in equal volumes of distilled water:
• use a range of temperatures
• leave for equal time.

↓

Remove pieces carefully and shake each solution gently to disperse any pigment.

↓

Assess amount of pigment lost using a colorimeter to measure the absorbance or transmission value of the solution.

↓

Plot values on graph:
• temperature on x-axis
• absorbance or transmission on y-axis.

There is little or no pigment leakage between these temperatures because the lipid layer in the membrane is intact. (1)

There is a sudden increase in leakage of pigment out of the beetroot pieces; (1) this is probably due to disruption of the membrane as it becomes more fluid at about 55 °C. Denaturation of membrane protein channels may also play a part. (1)

Graph showing typical results of an experiment to investigate the effect of temperature on the leakage of betalain pigment from beetroot tissue.

How are gas exchange surfaces adapted to allow rapid exchange?

Gas exchange surfaces rely on diffusion of gases across them and diffusion is more rapid:
• the larger the area
• the thinner the diffusion distance
• the steeper the concentration gradient.

In the lungs of humans, the rate of diffusion is maximised by adaptations of the gas exchange surfaces:

Factor	Adaptation of lung
surface area	Many alveoli give a huge surface area (often described as that of a tennis court).
concentration gradient	Well supplied with constantly circulating blood which carries oxygen away and carbon dioxide to the surface. Breathing keeps oxygen concentration high and carbon dioxide concentration low.
thickness of gas exchange surface	Very thin – just one flattened epithelial cell in thickness – reducing the distance for diffusion.

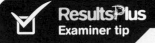

ResultsPlus
Examiner tip

Remember to describe *all* the features of a graph, not just the straight line parts, and only explain if you are asked to do so!

Worked Example

For the graph above, describe and explain the effect of increasing temperature on the permeability of the membrane of beetroot cells.

Description (2 marks): As the temperature increases from 0 to 40 °C there is little effect on pigment leakage (1 mark), then at about 50 °C the membrane becomes very leaky to pigment (1 mark).

Explanation (3 marks) is on the diagram.

Thinking Task

Q1 A red blood cell is placed in a solution which is more concentrated than the solution inside it. Draw a diagram to show the movement of water between cell and solution, using arrows to show direction and relative length to show how much.

Quick Questions

Q1 Which methods of movement across cell membranes require energy from ATP?
Q2 List all the passive methods of movement across membranes.

Structure and role of DNA and RNA

DNA and RNA are polynucleotides formed from subunits called **nucleotides**.

Structure of nucleotides

Nucleotides contain:
- a phosphate group
- a base: adenine (A), guanine (G), cytosine (C), thymine (T) in DNA; A, G, C and uracil (U) in RNA
- a sugar: ribose in RNA, deoxyribose in DNA.

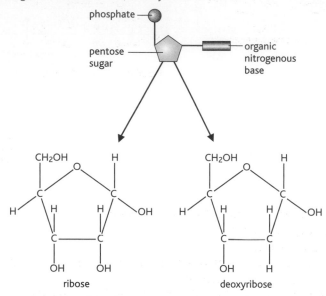

The structure of a nucleotide.

The components are joined in condensation reactions with the loss of water, the phosphate joins to carbon 5 of the sugar and the base joins to carbon 1.

Nucleotides join in long chains of theoretically infinite length by bonds formed in condensation reactions.

RNA is a single strand of nucleotides. In DNA, two separate strands are held together by hydrogen bonds. These strands wind around each other in a **double helix**. There is **complementary base pairing** which follows strict rules – A always pairs with T and G always with C. This is because of:
- size: A and G are large (2-ring molecules), C and T are smaller (1-ring molecules). One large base pairs with a smaller base so the width of a base pair is always the same.
- hydrogen bonding: due to the shape of the molecules, two H bonds form between A and T, whereas three form between C and G. Alternative pairs are not possible.

Examiner tip

When talking about base pairing, remember it is *complementary* base pairing and that adenine pairs with thymine not thiamine! The smaller bases with a Y in the name (C and T) are pYrimidines and bind to one larger purine (G or A).

Watch out!

It is very easy to muddle up features of DNA and RNA, so learn them well.

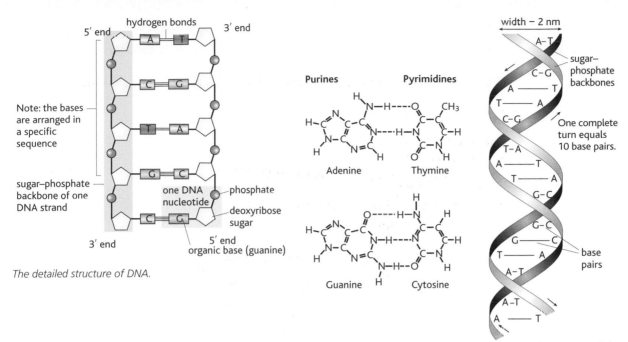

The detailed structure of DNA.

The two strands are antiparallel – one runs in one direction and the other in the opposite direction.

The gene and the genetic code

- The order of bases on one strand of DNA is the **genetic code**.
- This code is formed from **triplets** of bases.
- Each triplet codes for an amino acid.
- The sequence of triplets codes for the sequence of amino acids that will form a polypeptide, which will fold up to form a protein.
- A **gene** is a sequence of bases on one of the strands of a DNA double helix molecule which codes for a polypeptide chain (that is, a chain of amino acids).

DNA strand transcribed into mRNA and translated into amino acid sequence

triplet of bases in DNA strand

transfer RNA brings amino acid to mRNA molecule for translation

transfer RNA brings amino acid to mRNA molecule for translation

transfer RNA brings amino acid to mRNA molecule for translation

transfer RNA brings amino acid to mRNA molecule for translation

Different triplets code for different amino acids.

Relationship between genetic code and amino acid sequence. Note: for more details on transcription and translation see page 37.

Quick Questions

Q1 DNA consists of two strands held together by hydrogen bonds between complementary bases.
   a Name the base that pairs with thymine.
   b Name the base that replaces thymine in RNA.

Q2 Apart from bases, what other chemical groups occur in DNA?

ResultsPlus
Examiner tip

Do not try to learn the genetic code, but be prepared to be given it or maybe part of it in a question. You will then have to apply your understanding of the code.

Thinking Task

Q1 Make a table to compare the structure of DNA and RNA.

DNA replication and protein synthesis

The genetic code must:
- self replicate so that copies can pass to daughter cells during cell division
- carry information that codes for proteins.

When DNA was confirmed as the genetic code, there were competing theories about how it replicated. Meselson and Stahl's classic experiment with *Escherichia coli* showed that the **semi-conservative theory** was correct.

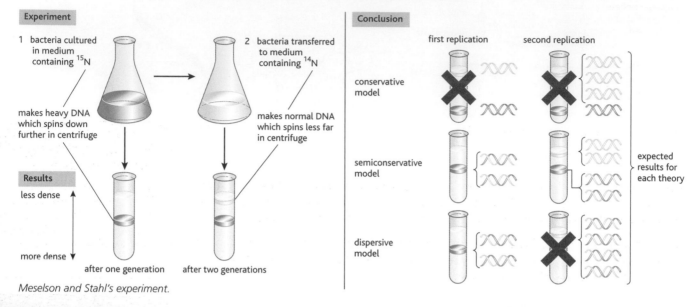

Meselson and Stahl's experiment.

How does DNA replication occur?

During replication the two strands of DNA unwind and split apart. Free nucleotides line up along each strand, observing the complementary base pairing rules (page 34). The enzyme DNA polymerase bonds the nucleotides together as a phosphodiester bond forms between each deoxyribose and adjacent phosphate group. Hydrogen bonding links the two strands together.

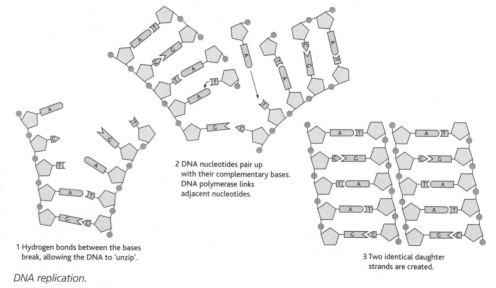

1 Hydrogen bonds between the bases break, allowing the DNA to 'unzip'.

2 DNA nucleotides pair up with their complementary bases. DNA polymerase links adjacent nucleotides.

3 Two identical daughter strands are created.

DNA replication.

Protein synthesis

The genetic code in DNA is in the nucleus, but the proteins formed using the code are made in the cytoplasm. So the DNA code is copied, making a molecule of **messenger RNA** in a process called **transcription**. The mRNA passes into the cytoplasm through nuclear pores and is used to make a polypeptide in a process called **translation**.

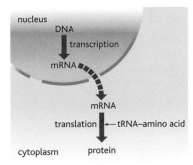

Summary of the steps in protein synthesis.

How does transcription work?

During transcription the DNA unwinds and hydrogen bonds between base pairs split to separate the two strands. Only one strand is used in the formation of mRNA – the **template** (**antisense**) strand. The unused strand is called the *sense* strand.

Ribonucleotides are paired with their complement on the template strand: uracil pairs with adenine instead of thymine. The ribonucleotides are then joined up by **RNA polymerase** to form a strand of mRNA.

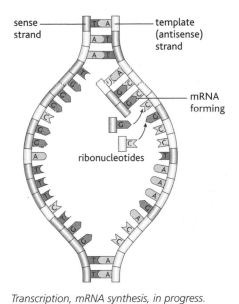

Transcription, mRNA synthesis, in progress.

What is translation?

The mRNA carries the genetic message in the same base sequence language as the DNA. Transfer RNA (tRNA) translates the base sequence on the mRNA into the protein amino acid sequence. Each tRNA molecule carries an amino acid to the mRNA, where the amino acid joins others carried by other tRNAs to build a polypeptide.

ResultsPlus
Watch out!

Make sure you have got the terms 'sense strand', 'antisense strand', 'template', 'codon' and 'anticodon' clear in your mind. Make a card with descriptions of each one on it.

ResultsPlus
Examiner tip

The words 'transcription' and 'translation' are very similar and very easily muddled. Try to remember this. When you go from English to French you are going from one language to another; you are translating. When you go from a base sequence (in mRNA) to an amino acid sequence in a polypeptide you are going from one language to another, so again translating. Once you have got this, you should also know what transcription is.

Quick Questions

Q1 Given the following sequence of bases on a strand of DNA, what would be the sequence on the complementary strand and on the mRNA molecule formed from it?

TACGGTATGCCAACCTTC

Q2 Write a definition of each of the following terms, in the context of DNA replication and protein synthesis:

transcription   translation   template strand   sense strand

Thinking Task

Q1 If *one* strand of DNA, during replication, has had ACT added to it, what will be the next nucleotide added – A, C, T or G, or can you not say? Explain your answer.

Q2 Make a table to compare DNA replication with transcription.

Genes, mutation and cystic fibrosis

At any stage when the genetic code is copied, mistakes can be made in the new base sequence formed. These mistakes are called **mutations**.

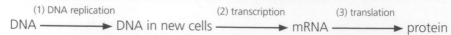

DNA $\xrightarrow{\text{(1) DNA replication}}$ DNA in new cells $\xrightarrow{\text{(2) transcription}}$ mRNA $\xrightarrow{\text{(3) translation}}$ protein

Mutations that occur during DNA replication can have the greatest effect because they are passed to new cells. In body cells, they may lead to cancer. In gametes, they can be passed to offspring and lead to genetic disorders such as cystic fibrosis (CF).

Mutation

A mutation is a change in a base sequence on the DNA, and this can give rise to a change in amino acid sequence in the protein.

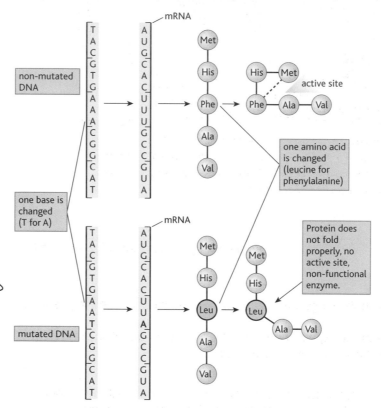

This type of mutation is called a substitution, where one base is substituted for another.

Cystic fibrosis

Cystic fibrosis (CF) is a genetic disorder caused by mutation of a single gene. This gene codes for the CFTR protein that allows chloride ions to pass through cell membranes. The gene is found on Chromosome 7. CF is one of the most common inherited conditions, with 1 in every 25 people in the UK carrying a recessive allele which causes it.

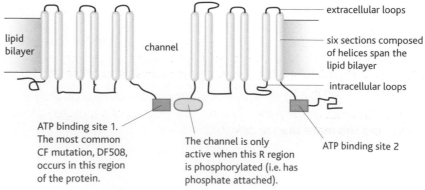

The CFTR protein is a channel protein.

Several different mutations in the CFTR gene can stop the protein working properly. One stops ATP binding to the CFTR protein, which stops the protein from opening to let chloride ions through. Another restricts chloride movement although the channel is open. The commonest mutation is due to the deletion (loss) of three bases which causes the loss of the 508th amino acid (phenylalanine); the finished protein cannot fold correctly to form the channel.

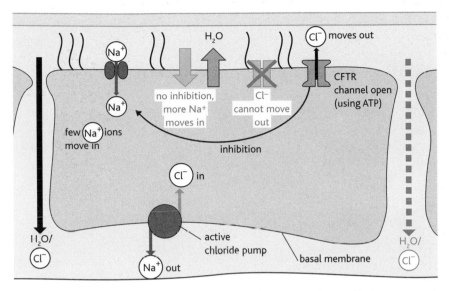

The normal functioning of an epithelial lining cell producing runny mucus (blue). Note: water leaves the cell at the top. The situation in a person with CF is shown in orange; in this case, water enters the cell at the top because Cl⁻ cannot leave to create the correct concentration gradient for water to move out by osmosis.

The effects of CF

The sticky mucus formed in CF can cause problems in gas exchange, digestion and reproduction.

Gas exchange:
- Mucus accumulates in the lungs, bacteria trapped in mucus increase the possibility of infection.
- Mucus can block bronchioles, which reduces the number of alveoli in contact with fresh air so reduces the surface area for gas exchange.

Digestion:
- Mucus blocks the pancreatic duct, so digestive enzymes can't reach the duodenum (small intestine) and food is not properly digested. This leads to tiredness and difficulty in gaining weight.
- Enzymes trapped within the pancreas cause fibrosed cysts and damage to insulin-producing cells, leading to diabetes.

Reproduction:
- In women, mucus can block the cervix preventing entry of sperm.
- In men, the vas deferens (sperm duct) is either missing or blocked with mucus, so sperm cannot leave the testes.

Quick Questions

Q1 Cystic fibrosis symptoms arise due to blockage of various tubes by mucus which is too sticky. Make a table to show which tubes are blocked and what problems this leads to.

Q2 What does CFTR not do properly when it is the product of a mutated allele, and why does this make mucus sticky?

Thinking Task

Q1 Using the DNA sequence in the diagram on page 38, draw out the consequences of:
   a loss of the third base from the top (C)
   b addition of a base (A) after the third from the top (C).

Q2 Make a table to show the effects of CF and how they are brought about.

Genetic inheritance

The monohybrid cross

A monohybrid cross is the inheritance of just *one* characteristic. There is one **gene** which controls height in pea plants, and it has two forms or **alleles**, one for tallness and one for dwarfness. The diagram shows a cross between a *tall* pea plant and a *dwarf* pea plant.

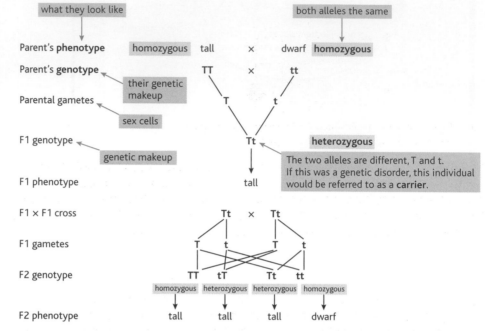

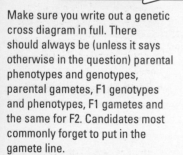

ResultsPlus
Examiner tip

If you are left to choose your own letters for the alleles of a cross, make sure you choose ones where the upper case is different from the lower (e.g. G and g, not O and o].

ResultsPlus
Watch out!

Make sure you write out a genetic cross diagram in full. There should always be (unless it says otherwise in the question) parental phenotypes and genotypes, parental gametes, F1 genotypes and phenotypes, F1 gametes and the same for F2. Candidates most commonly forget to put in the gamete line.

ResultsPlus
Examiner tip

The genetic cross diagram shows inheritance of height in peas, but you are expected to use the same kind of diagram to explain the inheritance of other monohybrid traits such as cystic fibrosis, albinism, thalassaemia, garden pea height and seed morphology (round, wrinkled, etc.). Punnet squares may also be used.

Because the dwarf allele is not **expressed** (seen) in individuals when it occurs with the tall allele, *tall* is said to be **dominant** and *dwarf* is said to be **recessive**. The convention for choosing the letter for the gene is to use the first letter of the dominant allele. In this case **tall** is dominant so we choose the letter t; uppercase, **T**, for the dominant allele, and lower case, **t**, for the recessive.

In human genetics, it is of course not ethically acceptable to set up crosses like the example with peas. The inheritance of characteristics in humans is studied by looking at natural crosses and the production of offspring using **genetic pedigree diagrams**.

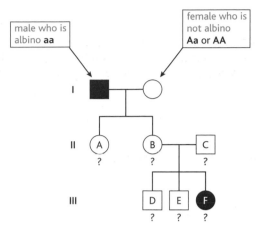

Pedigree diagram showing the inheritance of albinism (lack of pigment) in humans. The albino allele is recessive to the 'normal' one.

With a good understanding of the basic rules, it is possible to deduce a lot about the underlying genetics. In the diagram, F must be **aa** as albinism is recessive. B and C must be **Aa** as they have produced offspring F but they are both normal (they are carriers of albinism). The remaining individuals, D, E and A cannot be deduced beyond **Aa** *or* **AA** as the diagram does not have enough information about them.

ResultsPlus
Examiner tip

Make sure you write out genetic diagrams in full and with care. Many candidates hand in something which is very difficult to see any merit in because of the *unclear* use of *relevant* symbols and lack of a clear, logical structure.

Quick Questions

Q1 Make yourself a glossary of the following: gene, allele, genotype, phenotype, recessive, dominant, homozygote and heterozygote. These could be usefully written out on index cards so you can refer to them when you revise.

Q2 Suppose a woman carries cystic fibrosis and marries a man who does not. What is the chance that their child would inherit the disorder? Choose from the following options:

0%   25%   33%   50%   75%

Thinking Task

Q1 Use the pea height cross diagram on page 40 to draw a similar diagram for all the traits in the specification. Choose your own symbols based on the rules above.

- cystic fibrosis – the disorder is recessive
- albinism – the condition is recessive
- thalassaemia – the disorder is recessive
- seed morphology – wrinkled is recessive to round

(in peas Mendel studied wrinkled and smooth)

Gene therapy and genetic screening

Gene therapy is the insertion of a normal allele of a gene into cells to replace a faulty allele that causes an inherited disorder. Scientists are developing ways of using it to treat genetic disorders such as cystic fibrosis. This could be done in the very early embryo (**germ line therapy**) or in the affected body part or parts (**somatic therapy**).

Somatic therapy

- Step 1: identify the gene involved, for example the gene for CF is found on chromosome 7.
- Step 2: Make copies of the normal allele and insert into a **vector** – the two most common vectors for human cells are viruses and liposomes (spherical phospholipid bilayers).
- Step 3: Use the vector to insert the allele into the target cells.

Trials of gene therapy for CF have successfully transferred the normal CFTR allele to the epithelical cells in the lungs of CF patients. After insertion of the normal allele into the genome of the target cell it can make the functioning CFTR and therefore allow normal chloride movement. The mucus will be runny and the gaseous exchange system symptoms disappear.

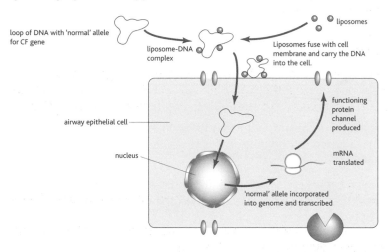

Inserting a 'normal' allele into a lung epithelial cell using a liposome.

Problems with current gene therapy for CF

- Only about 25% of normal chloride transport function is restored.
- Effect is temporary (up to 2 weeks) because epithelial cells are constantly replaced with new cells containing the faulty gene.
- Use of viruses can cause side effects: headache, fatigue, fever and raised heart rate.
- Delivery is very inefficient, especially with liposomes. Only about 1 in every 1000 genes gets into epithelial cells.

Avoidance and early treatment

Because no genetic disorder can be cured, potential parents have the following options:
- Avoid having a child with such a condition.
- Start treatment immediately after birth, which improves health in later years.

These options may involve **genetic screening**. Examples of screening include:
- Newborn babies could be tested for faulty alleles of the CF gene, but a blood test is routinely used to diagnose CF. The CF gene has many mutations and no test can cover all of them, so a negative result could be false.
- DNA testing of adults to identify carriers. A couple who were both carriers would have a 1 in 4 chance of having a CF baby (see page 40).

- Pre-natal DNA testing of embryonic cells using amniocentesis or chorionic villus sampling (CVS) so parents can decide whether to have the baby or not, and so that treatment can start immediately after the birth if the baby has CF.

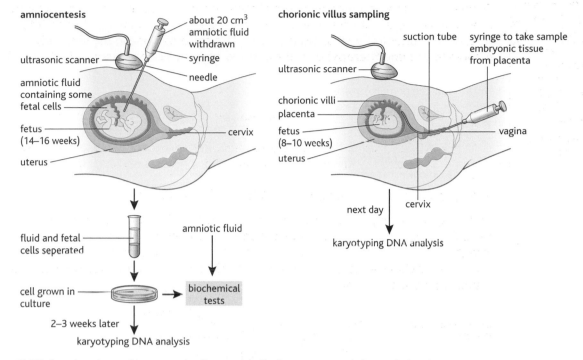

- PIGD (pre-implantation genetic diagnosis). Embryos created through *in vitro* fertilisation (IVF) are tested to see if they carry the faulty allele, only those which do not are implanted into the woman This method has the disadvantage that IVF is expensive and quite unreliable.

Making the right decision

There is no *right* decision on what to do about passing on a genetic disorder – different people have different attitudes. Ethical frameworks can help us to make the best decision.

- *Rights and duties* – Most people believe humans have rights to various basic things, and also that others have a duty to give them their rights.
- *Maximising the good (utilitarianism)* – This is a simple approach to life in which nothing is fundamentally wrong or right, but depends on the circumstances. The deciding principle is which action would lead to the most good – therefore the least bad (but, notice, not 'bad') would be favoured.
- *Making decisions for yourself* – The idea of signing a consent form before an operation stems from the belief that we can and should make our own decisions.
- *Leading a virtuous life* – Traditionally, there are seven virtues: justice, wisdom, courage, faith, hope, charity and moderation in all things.

Factors to be considered when deciding what is 'best' in relation to screening are:
- risk of miscarriage or harm to fetus
- right to life of the fetus
- abortion in the event of a positive diagnosis
- the cost of bringing up a baby that is 'disabled'
- mental and emotional issues surrounding the birth of a 'disabled' baby.
- being prepared for a baby born with CF or other genetic disorder

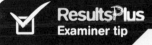

ResultsPlus
Examiner tip

Never make vague, unsupported statements in an examination, such as 'It is wrong to play God'.

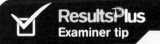

ResultsPlus
Examiner tip

Exam questions may expect answers that refer to a range of viewpoints.

⚙ Thinking Task

Q1 List *three* disadvantages of genetic screening.

Q2 Say what arguments you would put forward to *either* justify having an abortion *or* decide not to when you know you and your partner are both carriers of CF.

? Quick Questions

Q1 What do IVF, PIGD and CVS stand for?

Q2 Name two vectors for putting genes into human cells.

Topic 2 – Genes and health checklist

By the end of this topic you should be able to:

Revision spread	Checkpoints	Spec. point	Revised	Practice exam questions
Structure of amino acids and proteins	Describe the structure of amino acids and proteins and explain how the primary structure of a protein determines its three-dimensional structure and properties.	LO7	☐	☐
Enzyme action and rates of reaction	Explain the mechanisms of action and specificity in terms of 3D structure to explain how enzymes work as biological catalysts.	LO8	☐	☐
	Describe how enzyme concentration affects initial reaction rate. (core practical)	LO1, LO9	☐	☐
Membrane structure	Explain how the fluid mosaic model of membranes is supported by evidence.	LO2	☐	☐
Transport across cell membranes	Explain osmosis.	LO3	☐	☐
	Explain passive transport (diffusion and facilitated diffusion), active transport, endo- and exocytosis.	LO4	☐	☐
	Describe the investigation of the effect of temperature or alcohol on permeability of plant cell membranes. (core practical)	LO1, LO5	☐	☐
	Describe the properties of gas exchange surfaces, particularly the structure of the mammalian lung.	LO6	☐	☐
Structure and role of DNA and RNA	Describe the nature and role of a gene.	LO13	☐	☐
	Explain the genetic code.	LO12	☐	☐
	Describe the structure of mononucleotides and how they bond to form DNA (including complementary base pairing) and RNA.	LO10	☐	☐
DNA replication and protein synthesis	Describe DNA replication and the evidence for the mechanism.	LO11	☐	☐
	Outline protein synthesis, including transcription, translation, and the roles of tRNA and mRNA.	LO14	☐	☐
Genes, mutations and cystic fibrosis	Explain how errors in DNA replication give rise to mutations, such as that causing cystic fibrosis.	LO15	☐	☐
	Explain how the expression of a gene mutation in people with cystic fibrosis impairs function of the gas exchange, digestive and reproductive systems.	LO17	☐	☐
Genetic inheritance	Explain a range of important genetic terms used to describe aspects of monohybrid inheritance, and interpret pedigree diagrams.	LO16	☐	☐
Genetic screening and gene therapy	Describe gene therapy and distinguish between somatic and germ line therapy.	LO18	☐	☐
	Explain the use of genetic screening, and discuss the implications of prenatal genetic screening.	LO19	☐	☐
	Discuss social and ethical issues associated with genetic screening from a range of view points.	LO20	☐	☐

ResultsPlus
Build Better Answers

1 Cystic fibrosis is caused by a mutation in a gene coding for a membrane protein. Explain why people suffering from cystic fibrosis find it difficult to digest their food.

(4)

☑ Examiner tip

Always look at the number of marks available and the number of lines given for the answer. You do not need to fill the lines, but you do need to make enough points for the marks and if there are 8 or 10 lines, it is clear your answer will need to be substantial.

Student answer	Examiner comments
Food digestion takes place using enzymes in the small intestine. In a cystic fibrosis person these enzymes are either missing or there is not much of them. This is because the tube from the pancreas that carries the enzymes is blocked with mucus.	This answer makes two good points, one about enzymes being in short supply and one about this being due to a mucus blockage in the pancreatic duct. For 4 marks it would need to add something about why having fewer enzymes leads to a problem with digestion and/or why the duct has a mucus blockage in CF patients. ■ A **basic answer** would discuss the lack of enzymes from the pancreas or blockage of the duct. ▲ An **excellent answer** would mention the above and account for the mucus blockage in terms of its stickiness, itself due to the defective nature of CFTR leading to a reduction of chloride ions leaving the cell. It also might go on to talk about the low number of enzymes leading to fewer active sites to bind with substrates (proteins, polypeptides, lipids and starch).

2 The enzyme phenylalanine hydroxylase converts its substrate, phenylalanine, into the product, tyrosine. Using the information shown in the diagram and your knowledge of the mechanism of action of enzymes, suggest how this reaction takes place.

(4)

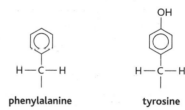

The structure of the R-group of phenylaline and tyrosine.

☑ Examiner tip

This question includes a specific enzyme and its substrate which you have not learnt about. The idea is that you use what you do know about enzyme action in this novel context. This will be a skill that will be tested frequently.

Student answer	Examiner comments
The phenylalanine enters the active site of the enzyme which is designed to fit phenylalanine only. The enzyme then catylises (candidate misspelling) the reaction by reducing the activation energy required to build it up into tyrosine.	This answer makes only two points for a 4 mark question. It gains a mark for the idea of the phenylalanine fitting into the active site and one for the ideas of lowering activation energy. ■ A **basic answer** would mention the idea of the active site and the substrate fitting into it, but would only get the mark if it was written in terms of phenylalanine and the specific enzyme. It would make no use of the diagram as asked for in the question. ▲ An **excellent answer** would discuss the formation of an enzyme–substrate complex between phenylalanine and phenylalanine hydroxylase, go on to talk about effects on activation energy or about bonds being broken, relate this to the specific reaction (addition of —OH to the phenylalanine R group) and say that the product tyrosine is then released from the active site because it can no longer bind to it, having changed.

Practice questions

1 (a) The diagram below shows a nucleotide with a nitrogenous base found in RNA but not DNA.

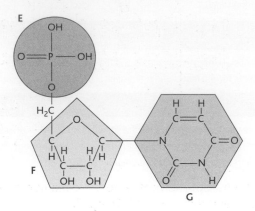

(i) Name the molecules labelled E, F and G. (3)

(ii) Name the part of the cell where RNA nucleotides are combined to form strands of messenger RNA (mRNA). (1)

(b) The table below shows which amino acids are coded for by different codons on mRNA.

First position	Second position				Third position
	U	C	A	G	
U	phe	ser	tyr	cys	U
	phe	ser	tyr	cys	C
	leu	ser	Stop	Stop	A
	leu	ser	Stop	trp	G
C	leu	pro	his	arg	U
	leu	pro	his	arg	C
	leu	pro	gln	arg	A
	leu	pro	gln	arg	G
A	ile	thr	asn	ser	U
	ile	thr	asn	ser	C
	ile	thr	lys	arg	A
	met	thr	lys	arg	G
G	val	ala	asp	gly	U
	val	ala	asp	gly	C
	val	ala	glu	gly	A
	val	ala	glu	gly	G

(i) The letters below represent a section of mRNA coding for the enzyme RNA polymerase. Using the table, give the amino acid sequence coded for by this mRNA sequence.

UACGUGGAAAGA (2)

(ii) Name the process that converts the mRNA sequence into a sequence of amino acids. (1)

(iii) In one body cell, a mutation occurs that changes the third base, cytosine, into guanine in this sequence of mRNA. Describe and suggest the potential effect this mutation could have on the protein (RNA polymerase) produced. (5)

(SNAB, 6131, AS Jun 2007, Q6)

2 **(a)** The diagram below shows a model of the cell membrane and various molecules being transported through the membrane into the cell.

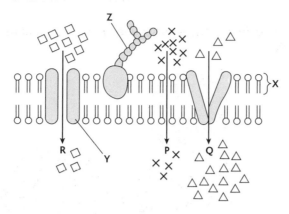

(i) Name the structures labelled X, Y and Z. (3)

(ii) Name the transport process across the membrane shown by the different molecules and their arrows P, Q and R. (3)

(b) Cystic fibrosis is caused by a mutation in a gene coding for a membrane protein. Explain why people suffering from cystic fibrosis find it difficult to digest their food. (4)

(SNAB, 6131, AS Jun 2007, Q4)

3 Achondroplasia is an inherited form of restricted growth in humans caused by a dominant allele. Individuals homozygous for the allele for achondroplasia are rarely born alive.

(a) Selecting suitable letters to represent the alleles, draw a genetic diagram to calculate the probability of a child inheriting achondroplasia if the mother is heterozygous for achondroplasia and the father has normal growth. (3)

(b) Two people who both have achondroplasia would like to have children together, but they are concerned about the risk of their child inheriting two achondroplasia alleles and dying before birth.

Name *one* method that could be used to obtain material suitable for use in a prenatal genetic screening program. (1)

(SNAB, 6131, AS Jun 2007, Q6)

4 One aim of gene therapy is to help overcome the effects of defective genes. A treatment for cystic fibrosis may become available in which copies of a functional gene are enclosed in small particles that can be sprayed into the lungs. The genes may enter the cells lining the lung and enable the cells to function normally.

(a) Explain how gene therapy could enable cells lining the lung to function normally in cystic fibrosis sufferers. (4)

(SNAB June 2005, AS Q4)

(b) If a couple have a child with cystic fibrosis, state the probability that a second child would unaffected. (1)

(c) What type of medical test could be offered to these parents if they were considering a second child? (1)

(SNAB January 2006, AS Q2 (modified))

Unit 1 Specimen paper

1 Read through the following passage that describes the process of blood clotting, then choose the most appropriate word or words to complete the passage where there are letters.

A blood clot may form when a blood vessel wall becomes damaged. Cell fragments called (A) stick to the wall of the damaged blood vessel forming a plug. A series of chemical changes occur in the blood, resulting in (B) being converted into thrombin. Thrombin is an (C) that catalyses the conversion of (D) into long insoluble strands of (E). These strands form a mesh that trap (F) to form the clot. (6)

Total 6 marks

(AS, 6BI01, Jan 2009, Q1)

2 Transcription and translation are two main stages in protein synthesis.

(a) Copy and complete the table below by writing the word *transcription* or *translation* next to the appropriate statement about protein synthesis. (5)

Statement	Stage of protein synthesis
ribosomes are involved	
DNA acts as a template	
tRNA is involved	
peptide bonds are made	
mRNA is made	

(b) The table below shows some amino acids and their corresponding DNA triplet codons. The DNA triplet codons for a stop signal are also shown.

Amino acid/stop signel	DNA triplet codons
proline	GGT GGG GGA
alanine	CGG CGA CGT CGC
cysteine	ACA ACG
serine	AGG AGA AGT AGC
leucine	GAA GAG GAT GAC
arginine	GCA GCG GCT GCC
glutamine	CTT CTC
glycine	CCT CCG CCA CCC
threonine	TGC TGA TGT TGG
stop signal	ATT ATC ACT

The diagram below shows part of a DNA molecule.

```
codon 84   codon 85   codon 86   codon 87   codon 88   codon 89
A C A C T T A C A G C C G G T G G G
```

(i) Write down the name of the amino acid that is coded for by codon 85. (1)

(ii) Write down the letter of the sequence of amino acids found in the polypeptide chain that is coded for by this part of the DNA strand.

A cysteine glutamine cysteine arginine proline proline
B threonine leucine threonine alanine glycine glycine
C cysteine glutamine cysteine arginine glycine glycine
D cysteine proline cysteine arginine proline proline (1)

(iii) If codon 89 coded for the last amino acid in the polypeptide chain, which letter was codon 90?

A GGG **B** ATC **C** TAG **D** AGT (1)

(iv) Write down the letter of the sequence of bases on a molecule of messenger RNA (mRNA) synthesised from this part of the DNA molecule.

A A C A C T T A C A G C C G G I G G G

B T G T G A A T G T C G G C C A C C C

C U G U G A A U G U C G G C C A C C C

D A G A C U U A G A C G G C C U G G G (1)

(v) Write down the letter of the statement that best describes what the polypeptide chain would be like if the 90th codon was ACT and the 91st codon was CTT on the DNA molecule.

A The polypeptide chain would be no more than 89 amino acids long.

B The 89th amino acid would be threonine and the 90th amino acid would be leucine.

C The polypeptide chain would be more than 90 amino acids long.

D The polypeptide chain would be more than 91 amino acids long. (1)

Total 10 marks

(AS, 6BI01, Jan 2009, Q2)

3 The diagram below represents the structure of the cell surface membrane.

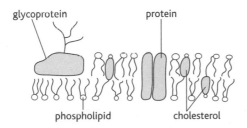

(a) Explain why the phospholipid molecules form a bilayer. (3)

(b) A student carried out an experiment to investigate the effect of alcohol concentration on the permeability of beetroot membranes. Beetroots are root vegetables that appear red because the vacuoles in their cells contain a water-soluble red pigment. This pigment cannot pass through membranes.

Eight pieces of beetroot were cut. One piece of beetroot was placed into a tube containing 15 cm³ of water and left for 15 minutes. The procedure was repeated for seven different concentrations of ethanol.

After 15 minutes, each piece of beetroot was removed from the tubes and a sample of the fluid removed and placed in a colorimeter. The colorimeter was used to determine the intensity of red coloration of the fluid.

The results of the investigation are shown in the graph below.

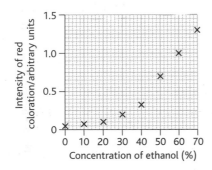

(i) Suggest *two* variables, other than those stated above, which should be kept constant during this experiment. (2)

(ii) There was some red coloration in the tube containing only water. Suggest an explanation for this. (2)

(iii) Describe what the student should have done to reduce the red coloration in the tube containing only water. (1)

(c) The graph above shows that ethanol has an effect on the permeability of beetroot.

   (i)   State the effect that the ethanol concentration has on the intensity of the red coloration. (1)

   (ii) Suggest an explanation for this effect. (2)

Total 11 marks
(AS, 6BI01, Jan 2009, Q4)

4 (a) Draw a labelled diagram to show the structure of an artery. (3)

   (b) Explain how the structure of an artery relates to its function. (2)

   (c) Give *two* differences between the structure of a vein and the structure of a capillary. (2)

Total 7 marks
(AS, 6BI01, Jan 2009, Q5)

5 High blood cholesterol levels are associated with an increased risk of developing cardiovascular disease (CVD). There are cholesterol-reducing drugs available to lower this risk.

   (a) Two groups of patients were treated with a different type of cholesterol-reducing drug, Drug A or Drug B.

   The graphs below show the percentage changes of total cholesterol (TC), low density lipoproteins (LDL) and high-density lipoproteins (HDL) in the blood of these patients, after treatment.

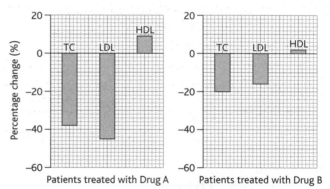

   (i)   Compare the effects of Drug A and Drug B on the percentage changes in total cholesterol (TC), LDL and HDL in the blood of these two groups of patients. (3)

   (ii) The enzyme HMG-CoA reductase catalyses the synthesis of cholesterol. When this enzyme is active, there are fewer LDL receptors on liver cells. These receptors are needed to remove LDL from the blood.

   Statins are a group of cholesterol-reducing drugs that act by inhibiting this enzyme. Suggest which of the two drugs, Drug A or Drug B, is more likely to be a statin. Give reasons for your answer. (3)

(b) State two risks of treatments using statins. (2)

(c) Age and gender are two other factors that may influence the development of heart disease in an individual.

The graph below shows the results of a survey in America, on the incidence of heart disease in adults aged 18 and older.

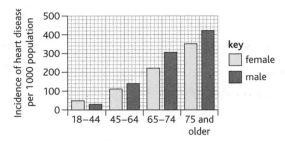

(i) Using the information in the graph, describe how the incidence of heart disease is affected by age and gender. (3)

(ii) Calculate the increased risk that a man who is 75 or older has of developing heart disease, compared to a man aged between 18 and 44 years old. (2)

Total 13 marks

(AS, 6BI01, Jan 2009, Q6)

6 DNA and lipids are important molecules found in living organisms. A triglyceride is one type of lipid. For each of the descriptions below, write down the letter of the correct statement about lipids or triglycerides.

(a) Triglycerides are composed of:

A three glycerol molecules and three fatty acid molecules.
B one glycerol molecule and three fatty acid molecules.
C one glycerol molecule and one fatty acid molecule.
D three glycerol molecules and one fatty acid molecule. (1)

(b) The bond between a glycerol molecule and a fatty acid molecule is:

A a glycosidic bond.
B a peptide bond.
C a phosphodiester bond.
D an ester bond. (1)

(c) This bond is formed by:

A hydrolysis.
B condensation.
C a chain reaction.
D an automatic reaction. (1)

(d) Unsaturated lipids:

A do not have any double bonds.
B have double bonds only between carbon atoms.
C have double bonds between carbon atoms and between carbon and oxygen atoms.
D have double bonds only between carbon and oxygen atoms. (1)

Total 4 marks

(AS, 6BI01, May 2009, Q1)

Cells and organelles

Some single-celled organisms and all multicellular organisms such as humans are made up of **eukaryotic** cells. A eukaryotic cell always contains:

- a nucleus, containing genetic material, surrounded by a double **membrane** (or envelope
- **organelles** (structures) in the cytoplasm, each surrounded by one or two membranes.

Animal cell ultrastructure

Ultrastructure is the name for the fine structure that is revealed when using a powerful microscope such as an electron microscope. The ultrastructure of a eukaryotic animal cell and details of some of its organelles are shown below.

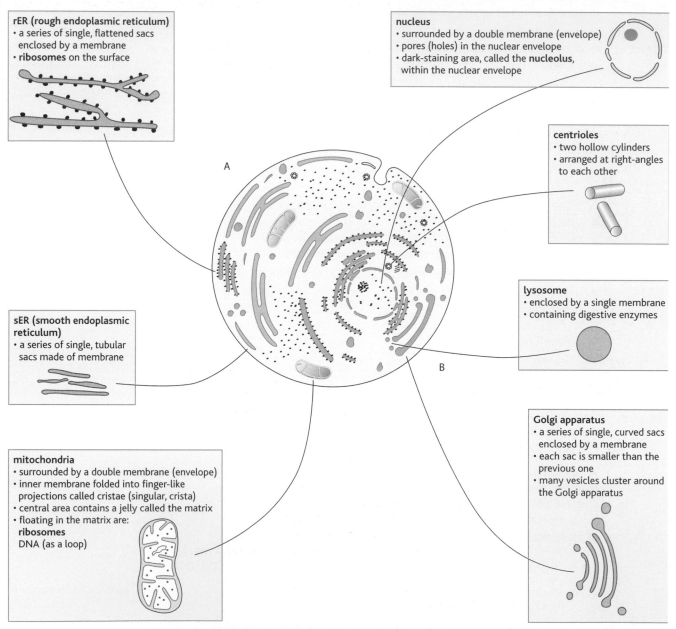

rER (rough endoplasmic reticulum)
- a series of single, flattened sacs enclosed by a membrane
- **ribosomes** on the surface

nucleus
- surrounded by a double membrane (envelope)
- pores (holes) in the nuclear envelope
- dark-staining area, called the **nucleolus**, within the nuclear envelope

centrioles
- two hollow cylinders
- arranged at right-angles to each other

sER (smooth endoplasmic reticulum)
- a series of single, tubular sacs made of membrane

lysosome
- enclosed by a single membrane
- containing digestive enzymes

mitochondria
- surrounded by a double membrane (envelope)
- inner membrane folded into finger-like projections called cristae (singular, crista)
- central area contains a jelly called the matrix
- floating in the matrix are:
  ribosomes
  DNA (as a loop)

Golgi apparatus
- a series of single, curved sacs enclosed by a membrane
- each sac is smaller than the previous one
- many vesicles cluster around the Golgi apparatus

Cross-section of a generalised animal cell – a typical eukaryotic cell.

Protein transport within eukaryotic cells

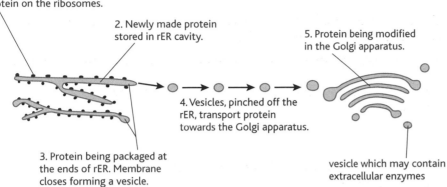

1. Amino acids made into protein on the ribosomes.

2. Newly made protein stored in rER cavity.

3. Protein being packaged at the ends of rER. Membrane closes forming a vesicle.

4. Vesicles, pinched off the rER, transport protein towards the Golgi apparatus.

5. Protein being modified in the Golgi apparatus.

vesicle which may contain extracellular enzymes

The Golgi apparatus and rER are involved in the making and movement of proteins through the cell. The arrows show the direction of transport. The proteins leave the cell by exocytosis, when the vesicles fuse with the cell surface membrane.

The modified protein is placed into vesicles by the Golgi apparatus. Some may be exported out of the cell, such as **extracellular enzymes**, e.g. amylase and protease.

ResultsPlus
Examiner tip

Make sure that you can recognise and label the organelles from electron microscope (EM) images as well as from diagrams.

Worked Example

One way to investigate protein trafficking is to use radioactive amino acids.

A tissue was soaked in a solution of radioactive amino acids for a short period of time and then transferred to a solution with non-radioactive amino acids.

The table below shows the amount of radioactivity, as a percentage of the total radioactivity in the cells of this tissue, found in three organelles, at three different times after being transferred to the non-radioactive solution.

Organelle	Percentage of total radioactivity in cells after tissue transferred to non-radioactive solution		
	5 min	10 min	45 min
vesicles	0	5	60
rER	80	10	5
Golgi apparatus	10	80	30
total radioactivity	90	95	95

(a) The vesicles containing radioactivity can either come from rER or the Golgi apparatus. Using evidence from the table, suggest whether these vesicles come from rER or the Golgi apparatus.

As most of the radioactivity is in the rER at 5 minutes, you would expect a lot in the vesicles soon afterwards if the rER made these vesicles. At 10 minutes, there was only a little in the vesicles and a lot in the Golgi apparatus. This suggests that the Golgi is just beginning to make these vesicles. This is confirmed at 45 minutes when there is less radioactivity in the Golgi apparatus as it is now in the vesicles. The vesicles, therefore, come from the Golgi apparatus.

(b) Suggest an explanation for the difference in the total radioactivity between 5 minutes and 10 minutes.

At first, it looks as though radioactive amino acids have suddenly appeared but this cannot be. This is a question about experimental technique. Data was collected at three time intervals rather than continuously. It could be that at 4 minutes more radioactivity was at the rER but by 5 minutes it has already been packaged and is in transit to the Golgi. By 10 minutes it has arrived at the Golgi.

Differences between a prokaryotic cell and a eukaryotic cell

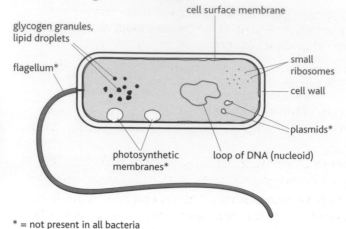

* = not present in all bacteria

A typical prokaryotic cell. Compare this to the diagram of the eukaryotic cell.

Feature	Prokaryotic cell	Eukaryotic	
		Animal cell	Plant cell
nucleus with envelope	absent	present	present
membrane-bound organelles	absent	present	present
DNA found as	a loop	separate strands	separate strands
slime capsule	sometimes present	never present	never present
flagella	simple (if present)	complex (if present)	absent
cell wall	present (bacterial)	absent	present (cellulose)
relative size of cell	small	medium	large

Table to compare a prokaryotic cell with two typical eukaryotic cells.

❓ Quick Questions

Q1 Give *three* structural differences between a typical prokaryotic cell and a human liver cell.

Q2 Give *one* similarity between the structure of a nucleus and a mitochondrion.

Q3 Describe how you could distinguish rER from Golgi apparatus.

Q4 List *four* membrane-bound organelles found in an animal cell.

⚙ Thinking Task

Q1 During cell division, an animal cell does not have a nucleus but is still considered to be eukaryotic. Suggest *two* reasons why this cell is still considered to be eukaryotic.

Q2 Draw a diagram of a Golgi apparatus.

Q3 Make a flowchart to describe protein trafficking from when the protein is first formed until it is released from the cell as an extracellular enzyme.

Cellular organisation

Multicellular organisms are organisms made up of many cells. These cells are not random throughout the body but are organised. There are *four* levels of organisation as shown below. Plant and animal examples are given.

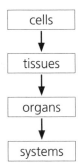

All **cells** of multicellular organisms are eukaryotic. Palisade cell and liver cell are examples.

Tissues consist of one or a few different types of cell that work together to perform a function. Epidermal tissue and muscle tissue are examples.

Organs are made up of various tissues grouped to work together and perform their function efficiently. Leaf and heart are examples.

Systems comprise various organs that work together to perform a large-scale function. Many people believe there are no systems in plants. The digestive system in humans is an example.

Worked Example

The diagram shows a section through a leaf. Explain why this leaf is considered to be an organ.

- The question asks about *this* leaf, so we must refer to the diagram. Note that there is no reference to chloroplasts so no marks would be given for writing about function.

- State that an organ is made up of several tissues. Point out that the diagram shows palisade tissue and also epidermal and spongy tissue. So this leaf has three tissues visible, hence is an organ.

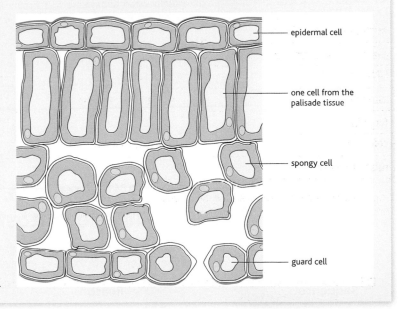

Section through a leaf.

epidermal cell

one cell from the palisade tissue

spongy cell

guard cell

Magnification and estimating size

Imagine a cell that measures 10 mm across on a page because it has been magnified 1000 times. From this we can work out the true size of the cell by using the formula below and the example.

Calculation stages

1 Using the pyramid, you can see that true size is found by dividing image size by magnification.

2 10×10^{-3} m (10 mm) ÷ 1000 =

3 The answer comes out to be 10×10^{-6} m or 10 µm

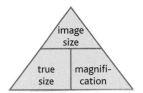

image size

true size | magnification

NB: To work out magnification using the pyramid, use image size ÷ true size.

Quick Questions

Q1 Explain what is meant by the term *system*.

Q2 Suggest *one* reason why organs are considered more complex than cells.

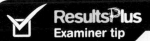

ResultsPlus
Examiner tip

Make sure all dimensions are in the same units before you start the calculation. This is a common source of error.

Thinking Task

Q1 Make a table to compare a tissue and an organ.

ResultsPlus
Examiner tip

When asked to compare, you should include similarities and differences.

The cell cycle and mitosis

The **cell cycle** and **mitosis** allow **asexual reproduction** (reproduction which involves only one individual) as well as growth in multicellular organisms. One cell undergoes division to form two cells, these two divide to form four cells and so on. This is **exponential growth**. All these cells have the same amount of DNA and there is no genetic variation.

The cell cycle is the sequence of events from:
- the formation of that cell
- until it divides to form daughter cells
- and includes three stages called interphase, **mitosis** and cytokinesis.

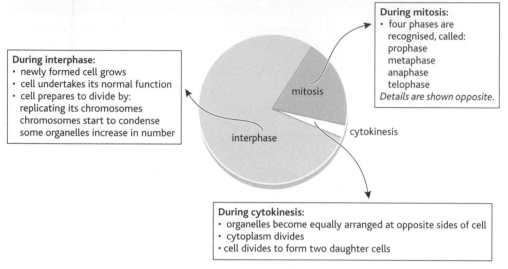

During interphase:
- newly formed cell grows
- cell undertakes its normal function
- cell prepares to divide by:
  replicating its chromosomes
  chromosomes start to condense
  some organelles increase in number

During mitosis:
- four phases are recognised, called:
  prophase
  metaphase
  anaphase
  telophase
Details are shown opposite.

During cytokinesis:
- organelles become equally arranged at opposite sides of cell
- cytoplasm divides
- cell divides to form two daughter cells

Pie chart to show approximate relative time a cell spends in each stage.

The root tip squash, a technique to view the cell cycle and mitosis

This is a core practical used to observe mitosis. You need to know the practical, why each stage is done and the risks involved in the practical.

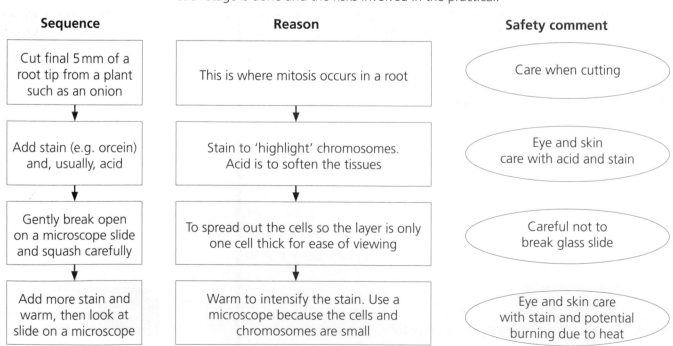

Sequence	Reason	Safety comment
Cut final 5 mm of a root tip from a plant such as an onion	This is where mitosis occurs in a root	Care when cutting
Add stain (e.g. orcein) and, usually, acid	Stain to 'highlight' chromosomes. Acid is to soften the tissues	Eye and skin care with acid and stain
Gently break open on a microscope slide and squash carefully	To spread out the cells so the layer is only one cell thick for ease of viewing	Careful not to break glass slide
Add more stain and warm, then look at slide on a microscope	Warm to intensify the stain. Use a microscope because the cells and chromosomes are small	Eye and skin care with stain and potential burning due to heat

Details of the four phases of mitosis

Phase	Main features	Image
prophase	• chromosomes continue to condense (and become visible) • nuclear envelope breaks down • nucleolus disappears • spindle forms • centrioles move to opposite poles of the cell (only in animal cells)	
metaphase	• chromosomes (as paired chromatids) line up at the cell equator (middle) • chromosomes attach to spindle fibres • fibres attach at the centromere	
anaphase	• spindle fibres contract • fibres pull chromatids apart with the centromere leading	
telophase	• chromosomes decondense (and become invisible) • nuclear envelope reforms • nucleolus reappears	

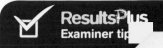

ResultsPlus
Examiner tip

Remember that mitosis is a continuous process so a cell at the start of prophase may not show all the points given in the table while a cell at the end of prophase does.

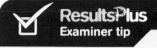

ResultsPlus
Examiner tip

Make sure you know the difference between centromeres and centrioles, and chromosomes and chromatids.

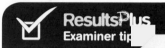

ResultsPlus
Examiner tip

Make sure you can identify the phases of mitosis from photographs as well as from diagrams.

? Quick Questions

Q1 Describe the events that occur during metaphase.

Q2 Make a flowchart to illustrate the term *cell cycle*.

Q3 Explain why a named stain and an acid are used during the root tip squash.

Q4 Using the pie chart at the start of this section, suggest which stage takes the longest amount of time. Give a reason for your answer.

Thinking Task

Q1 Describe and explain the role of the spindle during mitosis.

Production of gametes

Meiosis is a form of nuclear division whose functions are:

- production of **gametes** (sex cells such as sperm and ovum (egg) in animals); each gamete has half the number of chromosomes (haploid number, 23 in humans) found in a body cell
- to allow genetic variation to occur.

How genetic variation is achieved

1 Crossing over

During the first division in meiosis, pairs of chromosomes, known as homologous chromosomes, line up and may swap part of their genetic material.

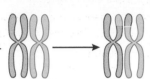

Homologous pair of chromosomes before crossing over.

Homologous pair after crossing over. All four components (chromatids) are now different from each other as two have swapped some alleles (genetic information).

2 Independent assortment

During the first division of meiosis, each homologous pair of chromosomes can line up as shown.

 or

Since each of the 23 pairs of homologous chromosomes in humans can line up either way round, it is highly likely that daughter cells will end up with different chromosome combinations when the homologous chromosomes separate.

Mammalian gametes are specialised for their functions

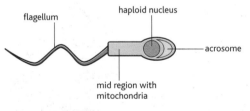

flagellum

haploid nucleus

acrosome

mid region with mitochondria

Sperm cell.

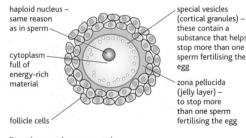

haploid nucleus – same reason as in sperm

cytoplasm full of energy-rich material

follicle cells

special vesicles (cortical granules) – these contain a substance that helps stop more than one sperm fertilising the egg

zona pellucida (jelly layer) – to stop more than one sperm fertilising the egg

Egg (secondary oocyte).

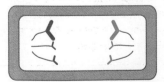
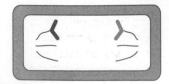
Worked Example

Describe and explain how a sperm cell is adapted for fertilising an egg cell.

Work logically; for example, start at the tip and work your way along the sperm cell.
- Acrosome contains digestive enzymes to enable sperm to penetrate zona pellucida (jelly layer).
- Nucleus contains the haploid number of chromosomes so that full complement restored at fertilisation.
- Mitochondria in mid region to provide useable energy to 'drive' the flagellum.
- Flagellum which whips back and forth to move sperm along towards ovum.

(?) Quick Questions

Q1 Make a table to show *four* structural similarities between a sperm cell and an egg cell.

Fertilisation

Fertilisation is the fusion of gametes. Mammalian gametes are sperm and ovum while pollen and egg are the gametes in flowering plants.

The importance of fertilisation in sexual reproduction is twofold:
- it restores the full complement (**diploid** number) of chromosomes (46 in humans)
- it allows genetic variation.

Fertilisation in mammals

Stage	Description of events
acrosome reaction	Acrosome releases digestive enzymes when sperm head meets the zona pellucida (jelly layer) of egg. Enzymes digest a channel in zona pellucida for sperm to burrow through to the cell surface membrane of egg cell.
membranes fuse	Cell surface membrane of sperm and egg fuse enabling haploid nucleus from sperm to enter cytoplasm of egg cell.
egg cell response (cortical reaction)	Special vesicles (cortical granules) move towards and fuse with the cell surface membrane. They release their contents (through exocytosis), which cause changes in surface layers of egg that stop other sperm from entering the egg cell.
meiosis restarted	The presence of the sperm nucleus in the cytoplasm of the egg cell causes the second division of meiosis to occur.
fertilisation	The chromosomes from the haploid sperm nucleus and from the haploid egg nucleus combine to restore the full complement of chromosomes, the diploid number.

Sequence of events ↓

Fertilisation in flowering plants

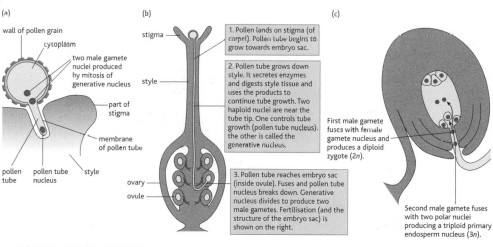

(a)
wall of pollen grain
cytoplasm
two male gamete nuclei produced by mitosis of generative nucleus
part of stigma
membrane of pollen tube
pollen tube
pollen tube nucleus
style

(b)
stigma
style
ovary
ovule

1. Pollen lands on stigma (of carpel). Pollen tube begins to grow towards embryo sac.

2. Pollen tube grows down style. It secretes enzymes and digests style tissue and uses the products to continue tube growth. Two haploid nuclei are near the tube tip. One controls tube growth (pollen tube nucleus), the other is called the generative nucleus.

3. Pollen tube reaches embryo sac (inside ovule). Fuses and pollen tube nucleus breaks down. Generative nucleus divides to produce two male gametes. Fertilisation (and the structure of the embryo sac) is shown on the right.

(c)
First male gamete fuses with female gamete nucleus and produces a diploid zygote (2n).

Second male gamete fuses with two polar nuclei producing a triploid primary endosperm nucleus (3n).

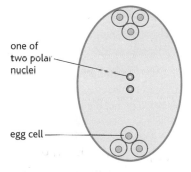
one of two polar nuclei
egg cell
Embryo sac.

Quick Questions

Q1 Describe what happens to the jelly layer during fertilisation.

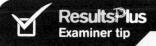

ResultsPlus
Examiner tip

Double fertilisation occurs in flowering plants, with:
- one male gamete fusing with an egg cell to form a diploid zygote
- one male gamete fusing with two polar nuclei to form a triploid primary endosperm nucleus that develops into the seed's storage tissue, endosperm.

Thinking Task

Q1 Using the diagram of the carpel (b), suggest the maximum number of seeds that could be formed. Give reasons for your answer.

Stem cells and cell specialisation

Stem cells are:
1 undifferentiated (unspecialised) cells
2 which can keep dividing
3 and that can give rise to other cell types.

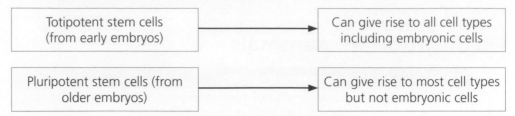

Totipotent stem cells (from early embryos)	→	Can give rise to all cell types including embryonic cells
Pluripotent stem cells (from older embryos)	→	Can give rise to most cell types but not embryonic cells

How to demonstrate totipotency

- Take a few plant cells of one type, called an explant.
- Place on agar which has certain growth regulators (growth hormones) added.
- Cells divide by mitosis to form a cluster of cells.
- Divide cluster and then place in containers with agar.
- Add different growth regulators to above to the agar which stimulate plant cells to differentiate into roots, stem and leaves, etc.

Using the totipotency core practical to illustrate How Science Works

ResultsPlus
Build Better Answers

As with all core practicals, you are expected to have a sufficiently good knowledge of the practical to tackle questions asked in unfamiliar ways (see Build Better Answers box, page 61).

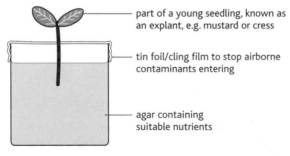

part of a young seedling, known as an explant, e.g. mustard or cress

tin foil/cling film to stop airborne contaminants entering

agar containing suitable nutrients

Example of equipment set up to show totipotency.

How Science Works

1 Risk analysis – working rapidly reduces the chances of an airborne organism entering, as does the presence of tin foil/cling film. Use agar with no sugar added so there is no food source present for any airborne organisms that do enter. Do not remove tin foil/cling film at end of practical.

⚠ Be careful if melting and pouring hot agar. Use gloves.

2 Evaluate methodology – one week later roots have formed. This example does not really show totipotency since only roots have formed.

How cells become specialised

Undifferentiated cells become differentiated cells when they develop a particular structure. This is due to differential gene expression.

Correct stimulus is given to the unspecialised cells, e.g. a chemical stimulus

↓

Some genes are switched on and become active; other genes are switched off

↓

Messenger RNA (mRNA) is made from the active genes only

↓

The mRNA moves to the ribosomes; the ribosomes read the mRNA and the appropriate protein is made

↓

The protein can permanently alter the structure and function of the cells

Examples of differentiated blood cells derived from bone marrow stem cells

erythrocyte (red blood cell)

leucocytes (white blood cells)

(?) Quick Questions

Q1 Give *two* similarities between a totipotent stem cell and a pluripotent stem cell.

Q2 Describe how mRNA helps to convert a pluripotent stem cell into a specialised cell.

Q3 Suggest why it is better to use several explants when trying to demonstrate totipotency rather than one explant.

Stem cells and medical therapies

Sources of stem cells

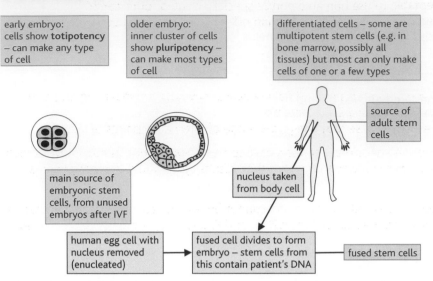

early embryo: cells show **totipotency** – can make any type of cell

older embryo: inner cluster of cells show **pluripotency** – can make most types of cell

differentiated cells – some are multipotent stem cells (e.g. in bone marrow, possibly all tissues) but most can only make cells of one or a few types

source of adult stem cells

main source of embryonic stem cells, from unused embryos after IVF

nucleus taken from body cell

human egg cell with nucleus removed (enucleated)

fused cell divides to form embryo – stem cells from this contain patient's DNA

fused stem cells

	Advantages	Disadvantages
embryonic stem cells	• easy to extract and grow	• ethical issues • possible rejection by patient's body • risk of infection when cells received • risk of stem cells becoming cancerous in body
adult stem cells	• fewer ethical issues • rejection risk avoided if stem cells taken from patient	• difficult to extract • more difficult to produce different cell types • risk of infection when cells extracted and received
fused cells	• rejection risk avoided if nucleus taken from patient • potential for treating genetic disorders	• ethical issues with source of embryonic nuclei • risk of infection when cells received • risk of stem cells becoming cancerous in body

Using stem cells

Since stem cells can form many different kinds of specialised cell, potentially they could be used to treat medical conditions where there is a loss, shortage or a reduced functioning of certain cell types. For example:

- Parkinson's disease – a progressive loss of nerve cells in the brain that are involved in muscle control
- multiple sclerosis (MS) – the electrical insulating layer surrounding nerve cells breaks down
- type 1 diabetes – caused when cells in the pancreas produce less than the normal level of insulin in response to a rise in blood glucose concentration
- burns – skin cells damaged so cannot be replaced.

Issues of using embryonic stem cells

- When does an embryo become a human with human rights?
  - Should there be a maximum age for embryos used in research?
  - Should human embryos be used at all?
- Is it acceptable to use human embryos specially created for research?
- Is it acceptable to fuse an adult human cell with a human egg cell to create new stem cells?

Who makes the decisions?

- people working in the stem-cell field because they have an understanding of the issues and what is and is not possible
- everyone else because they can give a range of alternative points of view.

To make informed decisions, the facts have to be known. Scientists must have carried out careful studies in a way that makes their results comparable to other people's research.

The final decisions about what can and cannot be researched are made by regulatory authorities such as the Human Fertilisation and Embryology Authority in the United Kingdom.

Quick Questions

Q1 There are risks associated with stem cell treatments. Give *two* possible risks to the person receiving the stem cell treatment.

Q2 Suggest why it may be easier to harvest pluripotent stem cells than totipotent stem cells.

Thinking Task

Q1 Various people are involved in making decisions about stem cell research. The table on the right gives *three* questions that require a decision. Place a tick (✔) in the box or boxes of the people that should be involved in making each decision.

Question	People working in stem cell research	Everyone else
Is the stem cell procedure technically possible?		
Is the stem cell procedure ethically acceptable?		
Should the maximum age for embryos used in research be increased (in the future)?		

ResultsPlus
Build Better Answers

The media is full of articles about the potential benefits of stem cell research.

(a) Suggest *one* benefit of using pluripotent stem cells rather than totipotent stem cells as a possible treatment for type 1 diabetes.

(b) Both totipotent stem cells and pluripotent stem cells come from embryos. One technique to produce such cells whilst reducing the ethical issues relating to the use of human embryos is by using a cow egg cell rather than a human egg cell. The nucleus is removed from the cow egg cell and is replaced by a human adult diploid nucleus to form an embryo. Suggest how this technique may reduce the issues relating to the use of human embryos, but will still produce human stem cells.

■ **Basic answer:**

(a) Candidate is likely to give a general benefit of using pluripotent stem cells rather than why it is better than a totipotent stem cell or why it is the best cell type for type 1 diabetes. Remember that it is a 'suggest' question which means that you are likely to be asked to apply your knowledge to an unfamiliar situation.

(b) It is likely that a reference would be made to the embryo not being human. Though a general link may be made to it being able to produce stem cells because it is an embryo it is unlikely that this will be developed to refer to humans (see good answer below).

▲ **Good answer:**

(a) A benefit to using pluripotent stems cells rather than totipotent ones, such as easier to obtain or more available, would be given.

(b) In addition to the embryo not being human, candidates are likely to comment that no human egg cell was used. However, the full complement of human genetic material is present in this embryo so human stem-cells can be produced.

NB You may consider the formation of hybrid cells unethical, but remember that this question is not asking about that.

Variation in phenotype

The **phenotype** is the outward expression of a cell or organism due to the interaction of the:

- **genotype**
- environment.

Relative effect of genotype and environment

It is very difficult to assess the relative contribution of each but studies showing the effect of each on their own can help.

1 Genotype

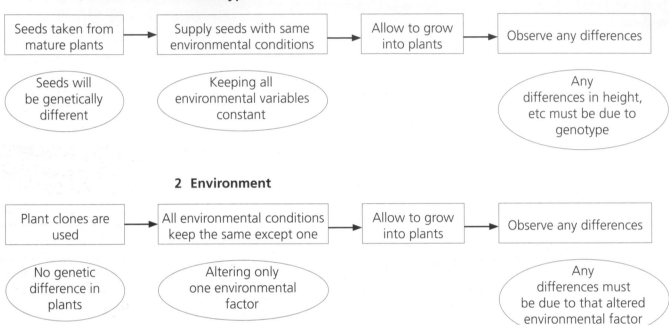

2 Environment

Examples of genotype and environment effecting phenotype

Siamese cat hair colour
- Genotype – gene codes for the enzyme tyrosinase which helps to make dark fur.
- Environment – enzyme only active when temperature drops below 31°C.
- Phenotype – distinctive colouration of Siamese cat because only body extremities, such as ears, drop below 31°C whilst rest of body is kept above 31°C.

Lung cancer and smoking
- Genotype – presence of proto-oncogenes involved in regulating the cell cycle (page 56).
- Environment – chemical components of cigarette smoke can alter these genes into oncogenes in lung cells.
- Phenotype – cell cycle may not be regulated by oncogenes in which case lung cells keep dividing without a check potentially leading to cancerous tumours in lungs.

Polygenic inheritance and continuous variation

In polygenic inheritance, more than one gene is involved in influencing the phenotype. The genes will be at different locations (gene loci) on the chromosomes. Variation in human height for a particular age group is an example.

Polygenic inheritance often gives rise to continuous variation where there are a few extremes and many in the middle.

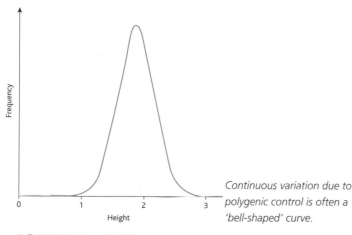

Continuous variation due to polygenic control is often a 'bell-shaped' curve.

Quick Questions

Q1 Human baby birth weight shows continuous variation.

a Suggest why baby birth weight shows continuous variation.

b Copy and complete the graph to show a suitable shape of curve for human baby weight.

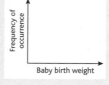

Q2 The maximum height of the flower stem of a particular alpine plant grown in optimum conditions was 10 cm. However, it was observed that in the wild, the maximum height was 2.6 cm. Suggest *three* environmental factors that could cause this height difference.

Thinking Task

Q1 The data in the table below suggest evidence that both genotype and environment have an effect on phenotype.

Situation in one twin	Likelihood of other twin also having type 1 diabetes (%)	
	Exposed to environmental trigger	Not exposed to environmental trigger
non-identical twin has type 1 diabetes	7.1	0.0
identical twin has type 1 diabetes	33.3	0.0

a State the phenotype.
b Give evidence that suggests genotype has an effect on the phenotype.
c Give evidence that suggests environment has an effect on the phenotype.

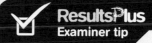

Topic 3 – The voice of the genome checklist

By the end of this topic you should be able to:

Revision spread	Checkpoints	Spec. point	Revised		Practice exam questions
Cells and organelles	Describe the structure of various organelles.	LO3		☐	☐
	Explain protein trafficking including the functions of rER and Golgi apparatus.	LO4		☐	☐
	Recall the structure of a typical prokaryotic cell. Be able to compare a typical eukaryotic cell with a typical prokaryotic cell.	LO2		☐	☐
Cell organisation	Describe how cells are organised into tissues, organs and systems.	LO5		☐	☐
The cell cycle and mitosis	Recall the meaning of the cell cycle and its three stages. Know the various phases of mitosis.	LO6		☐	☐
	Describe a practical to observe mitosis.	LO1, LO7		☐	☐
Production of gametes	Explain the functions of meiosis.	LO8		☐	☐
	Describe and explain the structure and function of human sperm and eggs.	LO9		☐	☐
Fertilisation	Describe the various stages in mammalian fertilisation. Describe fertilisation in flowering plants, including pollen tube formation.	LO10		☐	☐
Stem cells and cell specialisation	Explain the meaning of the terms totipotency and pluripotency and be able to compare the two.	LO11		☐	☐
	Explain how an unspecialised cell changes into a specialised cell due to differential gene expression.	LO13		☐	☐
	Describe a practical to illustrate totipotency in plants.	LO1, LO12		☐	☐
Stem calls and medical therapies	Recall the sources of totipotent and pluripotent stem cells. Discuss issues relating to the use of embryos to harvest these stem cells and their use in medical therapies. Discuss the need for regulatory authorities.	LO11		☐	☐
Variation in phenotype	Explain how a phenotype can be due to genotype and environment and that it can be hard to determine the relative contribution of each.	LO14		☐	☐
	Appreciate that some phenotypes are controlled by more than one gene and that those may show continuous variation.	LO15		☐	☐

ResultsPlus
Build Better Answers

1 Draw and fully label a diagram to show the structure of rough endoplasmic reticulum.

☑ Examiner tip

As you are being asked to draw and label, do not expect to gain all marks from just a good drawing. Make the diagram look like the structure. Not too small to see but not so big that you cannot add the labels. Labelling lines must touch the object you are labelling.

Student answer	Examiner comments
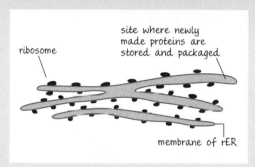	This is a good diagram and would gain all drawing marks. The ribosome label line does not touch the structure whilst the top right-hand label gives a function not structure as requested.
	■ A **basic answer** would include only a simple sketch, possibly with the ribosomes not attached to rER. It is likely that only one feature would be drawn and labelled. This would get a low number of marks.
	▲ An **excellent answer** would illustrate and correctly label all three key features, namely the presence of ribosomes, the rER membrane and its cavity.

2 The human body contains many different types of specialised cell, including gametes. A sperm cell is an example of a human gamete. The head of a sperm cell contains a haploid nucleus and an acrosome.

(a) Describe the structure of the human sperm nucleus.

☑ Examiner tip

It is possible that you will come across questions that are set in a novel way. It is unlikely that you will have been specifically taught the structure of a sperm cell nucleus. You should be able to use your knowledge and understanding from the course to tackle such questions.

Answer key	Examiner comments
Double membrane / envelope present; Nuclear pores present; Contains 23 chromosomes/haploid number of chromosomes;	As stated in the stem of the question, a sperm cell is an example of a human cell and is, therefore, a eukaryotic cell. The nucleus is likely to be structurally similar to other eukaryotic cells. However, there is half the full complement of chromosomes.

(b) The acrosome contains glycoproteins. Describe the trafficking of newly formed proteins until they have become glycoproteins.

☑ Examiner tip

Read questions very carefully. This one asks you to start with the newly formed protein, therefore, writing about ribosomes making the protein will score no marks. Likewise, giving details about what happens to the glycoprotein goes beyond the scope of the question and will not gain marks.

Answer key	Examiner comments
Protein enters the rER cavity; Protein packaged in the rER; Details of packaging; Vesicles formed move to the Golgi apparatus; Vesicles fuse with Golgi apparatus; As protein moves through Golgi apparatus it is modified to form glycoprotein;	A question such as this is likely to be worth several marks. Many good answers tend to be written in a sequential manner. An example would start with the newly formed protein entering the rER cavity where it becomes packaged into membrane-bound vesicles.

Practice questions

1 In the root tip of a plant there is a region of one type of cell whose function is to undergo mitosis.

  (a) **(i)** Name *one* organelle, visible using a light microscope, that suggests the cells in this region are eukaryotic. (1)

      (ii) Suggest the level of cellular organisation of the region described above. Give a reason for your answer. (2)

  (b) Below is a partially completed diagram of a plant cell in metaphase of mitosis. Copy and complete the diagram by adding in the arrangement of spindle fibres. (2)

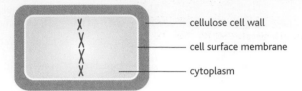

  (c) In this region, 10% of the cells are undergoing mitosis and 85% are in interphase. Name the stage that accounts for the remaining 5% of cells. (1)

  (d) Some of the cells from this region may become fibres. Describe how you could safely and reliably compare the tensile strength of plant fibres from two different species of plant to find which has the greater tensile strength. (4)

2 Stem cells are found in many areas of the human body and may be used as future medical therapies for a number of conditions.

  (a) Write down the letter that correctly applies to both totipotent and pluripotent stem cells. (1)

    A Can give rise to totipotent and to specialised cells.

    B Can give rise to pluripotent stem cells and to specialised cells.

    C Can give rise to all cell types.

    D Can give rise to only some cell types.

  (b) If a stem cell can divide every 24 hours, calculate the maximum number of stem cells there could be after 120 hours. Show your working. (2)

  (c) Explain how two identical stem cells could become two different specialised cells such as a cell from the brain and a cell from the pancreas. (3)

3 The phenotype of an organism may be affected by only the genotype, only the environment or a combination of both the genotype and the environment.

  (a) The table below describes four phenotypes. Copy the table and tick (✓) the box that best represents the factor that affects each phenotype. (4)

Description of phenotype	Genotype only	Environment only	Genotype and environment
a set of non-identical twins being a boy and a girl			
an arctic fox producing dark fur in summer and white fur in winter			
a human's lungs enlarge if they do a lot of swimming			
identical twins having different heights at the same age			

(b) The graph below shows an example of continuous variation due to the interaction of the environment and polygenic inheritance.

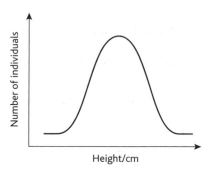

(i) What is meant by the term polygenic inheritance? (1)

(ii) The removal of outliers would alter the graph above. Write down the letter of the graph below that represents the example after the outliers have been removed. (1)

4 (a) In an investigation into pollen tube growth, the effect of two potential inhibitors, called A and B, was studied.

Pollen grains were allowed to grow in suitable conditions for 2 hours and then the lengths of their pollen tubes were measured. The mean pollen tube length was then calculated.

This was then repeated except that either substance A or substance B was present. The results are shown in the table below.

Conditions	Mean length of pollen tube after 2 hours/μm
suitable	450
suitable plus substance A	60
suitable plus substance B	300

(i) Describe the effect of substances A and B on mean pollen tube growth. (3)

(ii) Suggest *one* factor that needed to be kept constant in this investigation other than the time taken for pollen tube growth. (1)

(b) The mammalian sperm cell and the egg cell are adapted for their functions.

(i) Name *one* structure found in a sperm cell that is absent in an egg cell. (1)

(ii) Describe *two* responses of the egg cell when the sperm head enters the egg. (2)

Species and biodiversity

Biodiversity is the variety of different organisms within a habitat. Two aspects to consider are:

- Species richness – the number of different species within an area of known size at a particular time. A species is a group of organisms with so many features in common that they can mate and produce sexually viable offspring.
- Genetic diversity – the genetic variation within a species.

How to measure these two aspects of biodiversity

1 Measuring species richness

To estimate the species diversity, random sampling can be carried out. A number of quadrats (often 10) of known size are randomly placed in an area. All the species found in each quadrat are recorded and used to estimate the total species richness.

Worked Example

1 Using the data in the table:

  a work out the number of species found in the 3 m² sampled.

  b construct a bar chart to compare the frequency of occurrence of each species.

Answers

  a It doesn't matter whether the species occurs only once or many times as this question item is asking for presence or absence only. The answer is, therefore, 5 as A to E occur at least once.

	1 m² quadrat number		
Species	1	2	3
A	✔		
B			✔
C		✔	✔
D	✔	✔	✔
E	✔		✔
F			

  b Frequency is the total number of times each species is found. For species A it is 1 and for species D it is 3. The bar chart is shown below.

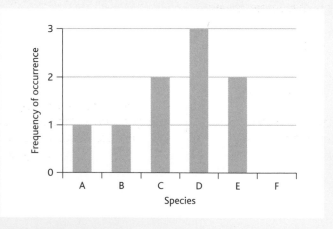

2 Measuring genetic diversity

Find the number of different alleles in a gene pool.

Each gene may have one to many different alleles.	The total number of genes a species has

Endemism

Endemism describes the situation where a species is found in only one particular area. The ring-tailed lemur is restricted to the island of Madagascar and so is considered to be endemic to Madagascar.

A new taxonomic grouping

Three domains have been identified based on **molecular phylogeny**. Molecular phylogeny compares the structure of a particular molecule from different organisms to discover their degree of evolutionary relatedness. The more similar the structure of the molecule, the more closely related the organisms are to each other. This is because changes in molecule structure generally occur only slowly as they are caused by mutations.

The three domains are:

- Bacteria
- Archaea
- Eukaryota.

The idea of the three domains illustrates the scientific process and the important role of critical evaluation of new data by the scientific community.

> **ResultsPlus**
> **Examiner tip**
>
> Mutations in DNA and the structure of the proteins produced are covered in Unit 1.

All life is split into the prokaryotes and eukaryotes	Most scientists accepted that all life was either prokaryotic (cells without a true nucleus) or eukaryotic (cells with a true nucleus).
A scientist called Woese suggested a third group which he called the Archaea; he supported his suggestion with evidence	Woese published a paper in a scientific journal using molecular phylogenic evidence. This is one way of informing the scientific community of a discovery. Other ways include presenting information at a conference or using the Internet.
Other scientists study the evidence carefully; this is called *peer review*	The scientists are checking that the evidence is accurate and correct, that the conclusions drawn are sensible and that the methods used are appropriate.
Other scientists start to collect evidence	The evidence will help to support the suggestion, to reject the suggestion or lead to a modification of the suggestion. There is still debate about Woese's suggestion but much of the scientific community now accept the three domains.

?) Quick Questions

Q1 Make a flowchart to explain how species richness can be measured.

Thinking Task

Q1 In a large population of organisms, eight different alleles were found for one gene. However, when only half of the population was sampled, six different alleles were found for the same gene. Suggest *two* possible reasons for the absence of the two alleles.

Natural selection and evolution

The **niche** of a species is the way that the species exploits its environment (e.g. a rabbit is a grassland herbivore). If two different species are present in the same niche at the same time, there will be **competition** and one will out-compete the other so that the better adapted will survive.

Types of adaptation

There are three main types of adaptation that improve the chances of survival. Most organisms show more than one type.

Type of Adaptation	Description	Examples
anatomical	A physical/ structural adaptation – it may be external or internal.	• Cacti have highly modified leaves called spines. The reduction in size decreases the number of pores (stoma) through which water can be lost to the air. • The rat-tailed maggot, which lives in water with a low oxygen concentration, has a long breathing tube that extends from its body to just above the water surface. This allows the animal to take oxygen from the air above the water. Atmospheric air contains more oxygen than water. • Kidneys of some desert mammals are modified (long loops of Henlé) to reduce water loss by producing very concentrated urine.
behavioural	A change in the behaviour of an organism to increase its survival chances.	• Some ectothermic (cold-blooded) organisms such as lizards orientate themselves to maximise their absorption of heat from the Sun until they reach their active temperature. • A single-celled organism reverses direction if it accidentally moves into unfavourable conditions from more favourable conditions.
physiological	These tend to be changes in the internal biochemical functioning of the organism in response to an altered environmental stimulus.	• People who move from sea level to high up a mountain slowly increase their oxygen-carrying capacity by producing more red blood cells. • Formation of hard skin on hands due to repeated pressure.

Natural selection can lead to adaptation and evolution

• Natural selection is the survival of individuals in a population because they have alleles that improve their chances of survival and reproduction.
• Evolution is the change in the frequency of certain alleles in a gene pool over time due to natural selection.

Evolutionary sequence

Bacterial example

| | cell without mutant enzyme |
| | cell with mutant enzyme |

A population of organisms shows genetic diversity due to having a variety of alleles (caused by mutations)

In a population of bacteria, some individuals have a mutant allele that produces an enzyme that can breakdown penicillin. Other bacteria do not have this enzyme. The enzyme has no particular effect on survival in normal conditions.

Environmental conditions change

Penicillin used as a drug to treat bacterial infection.

Natural selection removes some individuals with alleles that are not as advantageous

Bacteria that contained the allele for the enzyme are resistant to penicillin, and so are more likely to survive.

The remaining individuals grow and reproduce, passing on the advantageous alleles

Resistant bacteria reproduce and pass on the resistance allele to offspring.

Over many generations the frequency of these alleles increases

If penicillin continues to be used, eventually all the bacteria will have the allele for resistance.

Quick Questions

Q1 The table below describes two adaptations. Copy and complete the table by placing a tick in the box that represents the correct type of adaptation.

Description of adaptation	Anatomical	Behavioural	Physiological
bee orchid flowers look similar to insects			
a peahen is said to select her peacock mate on the size and shape of his tail			

Q2 Explain the meaning of the term *natural selection*.

Thinking Task

Q1 Copy and complete the table below by suggesting an environmental change that could cause the following changes in each of the individuals.

Description	Environmental change
a small UK mammal has extended its range north over the last 50 years	
tooth shape has become different in foxes that live in a city compared with foxes that live in the countryside	

Conservation and genetic diversity

There are numerous reasons why it is important that we conserve endangered species and two ways to do so are:
* zoos
* seed banks.

Illegal ivory product.

The roles of zoos in conserving endangered animals

1 Education
Informing all age groups who visit zoos of various conservation issues including:
* the illegal trade in certain animal products such as ivory
* the need to maintain biodiversity
* captive-breeding programmes including success stories.

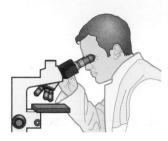

Scientific research.

2 Scientific research
Zoos, universities, etc. can work together on projects that are of benefit to the conservation of animals, including:
* control of diseases that are reducing populations
* behavioural studies to further appreciate the needs of animals in captivity
* development of techniques to further improve breeding success.

3 Captive-breeding programmes
These are schemes designed to encourage endangered species to breed so that:
* their numbers increase, reducing the risk of extinction
* subsequently some individuals can be released into the wild or protected areas such as national parks, to maintain or re-establish wild breeding populations
* the genetic diversity of the species is maintained.

The giant panda has been successfully bred in captivity.

4 Reintroduction programmes
These programmes endeavour to release captive-bred individuals back into the wild so that natural breeding populations can be maintained or re-established.

The importance of maintaining genetic diversity during captive breeding programmes

The problem is that some zoos can only have a small number of individuals, so inbreeding is likely to occur. This leads to:
* reduced genetic diversity and, therefore, a reduced chance of adapting to environmental change
* increased risk of a genetic condition becoming more common in the breeding population

The Arabian oryx is an example of reintroduction.

Some of the techniques used to reduce inbreeding and maintain genetic diversity
Do not allow the organisms to repeatedly breed with the same partner, possibly by isolating partners.
Select partner, possibly by adding a potential partner to a cage, IVF or inter-zoo swapping.
Keep a record/database of individuals in captivity and their breeding history, e.g. stud books, so that choice of partners is controlled,

The role of seed banks in conserving endangered plants

Seeds from a variety of endangered plants can be stored in a dormant state in seed banks (though not all seeds can be stored this way). Seeds rather than living plants are stored because:

- less space is required so more species can be held in the available space
- most plants produce large numbers of seeds so collecting small samples is unlikely to damage the wild population
- easier to store because dormant
- more cost effective.

How a seed bank can be used to conserve a rare plant species

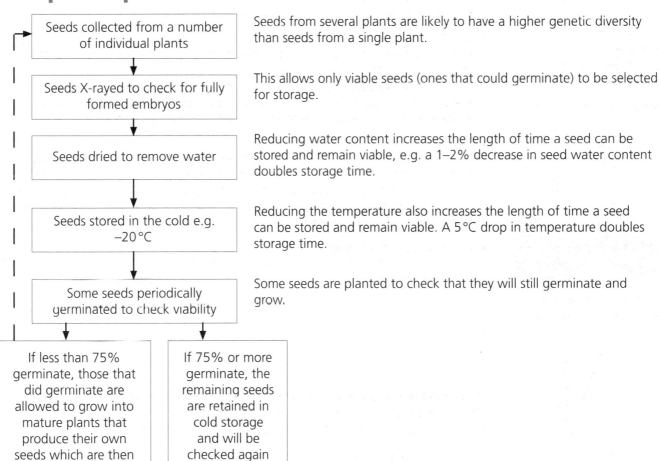

Seeds collected from a number of individual plants

Seeds from several plants are likely to have a higher genetic diversity than seeds from a single plant.

Seeds X-rayed to check for fully formed embryos

This allows only viable seeds (ones that could germinate) to be selected for storage.

Seeds dried to remove water

Reducing water content increases the length of time a seed can be stored and remain viable, e.g. a 1–2% decrease in seed water content doubles storage time.

Seeds stored in the cold e.g. –20 °C

Reducing the temperature also increases the length of time a seed can be stored and remain viable. A 5 °C drop in temperature doubles storage time.

Some seeds periodically germinated to check viability

Some seeds are planted to check that they will still germinate and grow.

If less than 75% germinate, those that did germinate are allowed to grow into mature plants that produce their own seeds which are then stored

If 75% or more germinate, the remaining seeds are retained in cold storage and will be checked again for viability

Quick Questions

Q1 Suggest how inter-zoo swapping will help to maintain genetic diversity.
Q2 Suggest why decreasing the temperature and water content increases the length of time seeds can remain in storage.

Plant cell structure

Plant cells are **eukaryotic** so have many features in common with animal cells such as:
* a nucleus surrounded by a double membrane (or envelope)
* organelles in the cytoplasm surrounded by one membrane (e.g. Golgi and rER) or two membranes (e.g. mitochondria).

The major ultrastructure differences are the additional structures present in plant cells.

1 Chloroplasts (in some cells e.g. palisade cells)

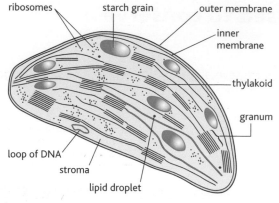

Diagram of a chloroplast.

Structural features:
* double membrane present
* many internal (thylakoid) membranes; some are arranged as stacks called grana (singular, granum)
* chlorophyll found on the thylakoid membranes
* fluid-filled interior called the stroma
* a loop of DNA is found in the stroma
* starch grains may also be present in the stroma.

2 Amyloplasts (in some cells, e.g. in some potato cells)

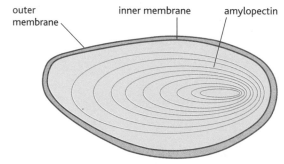

Diagram of amyloplast.

Structural features:
* double membrane present
* contains one type of starch called amylopectin
* amylopectin sometimes shows concentric rings.

3 Vacuole

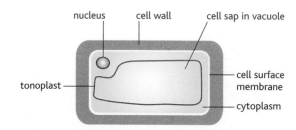

Diagram of a simple plant cell.

Structural features:
* surrounded by a single membrane called the **tonoplast**
* contains cell sap.

4 Cell wall

All plant cells have a cellulose cell wall, outside the cell surface membrane, which contains the following components:
- middle lamella – helps hold adjacent cells together
- plasmodesmata – involved in cell-to-cell transport
- pits – also involved in cell-to-cell transport.

Primary and secondary cell walls are described in more detail on page 80.

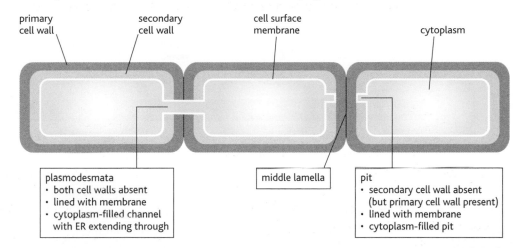

Comparison of the ultrastructure of a typical plant cell with a typical animal cell

ResultsPlus
Examiner tip

An example of a typical plant cell and animal cell may be given, such as root hair cell and liver cell.

Feature present	Typical plant cell	Typical animal cell
cellulose cell wall	✔	✘
plasmodesmata	✔	✘
pit	✔	✘
chloroplast	✔	✘
sap vacuole/tonoplast	✔	✘
cell surface membrane	✔	✔
mitochondria	✔	✔
larger (80s) ribosomes	✔	✔
rough endoplasmic reticulum	✔	✔
smooth endoplasmic reticulum	✔	✔
Golgi apparatus	✔	✔
nucleus	✔	✔
nucleolus	✔	✔
centrioles	✘	✔

Quick Questions

Q1 Give *two* similarities between plasmodesmata and pits

Q2 Make a table to compare the structure of a chloroplast and an amyloplast

Thinking Task

Q1 It has been hypothesised that chloroplasts and mitochondria have a similar evolutionary origin. Describe the structural aspects of these two organelles that would give supporting evidence for this hypothesis.

Plant stem structure and function

The plant stem contains a variety of structures including sclerenchyma fibres and xylem vessels.

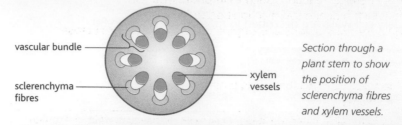

Section through a plant stem to show the position of sclerenchyma fibres and xylem vessels.

Sclerenchyma fibre	Xylem vessel
(diagram)	(diagram) *Cut section of vessel*
short structures with tapered ends	long cylinders (made up of a column of cells whose end walls have broken down)
ends closed	ends open
tough lignin present in walls	tough lignin present in walls

Table to show the structure of sclerenchyma fibres and xylem vessels.

Function	Sclerenchyma fibre	Xylem vessel
support (due to tough lignin)	✔	✔
water and mineral ion transport	✗	✔

Table to show the functions of sclerenchyma fibres and xylem vessels.

Worked Example

Describe and explain how the structure of sclerenchyma fibres and the structure of xylem vessels are appropriate to their functions.

ResultsPlus
Build Better Answers

Always read the question carefully. This one asks you to describe the relevant structure of sclerenchyma and xylem so there are marks for this. However, it also asks you to then link the structure to their functions. There will be marks for functions so long as they are correctly linked.

Answer

- For sclerenchyma, the appropriate structural reference relates to the deposition of lignin in the cell walls. The link to the function is that lignin is strong to aid in the support of the plant.

- For xylem, the two structural aspects are the presence of lignin and that they are long hollow tubes. This allows the two functions of structural support to the plant and a continuous pipe to allow water and mineral ions to move up the stem.

The importance of water and inorganic ions to plants

Plants need water for:
* inflating cells (turgor) which helps to keep plants upright
* as a component of chemical reactions such as in photosynthesis
* as the medium in which biochemical reactions occur in cells
* as a transport liquid for mineral ions and organic molecules such as sucrose.

Mineral ion	Importance
nitrate	to supply nitrogen for making proteins, DNA, RNA and certain plant hormones
calcium	to form calcium pectate in plant cell walls; to be involved in membrane permeability
magnesium	as a component of chlorophyll; to help in the formation of DNA; as an activator of certain plant enzymes

Table to show the importance of some mineral ions to plants.

Mineral ion deficiency practical

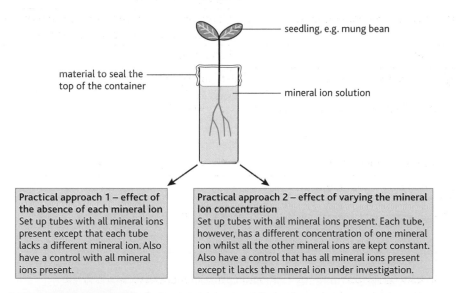

seedling, e.g. mung bean

material to seal the top of the container

mineral ion solution

Practical approach 1 – effect of the absence of each mineral ion
Set up tubes with all mineral ions present except that each tube lacks a different mineral ion. Also have a control with all mineral ions present.

Practical approach 2 – effect of varying the mineral ion concentration
Set up tubes with all mineral ions present. Each tube, however, has a different concentration of one mineral ion whilst all the other mineral ions are kept constant. Also have a control that has all mineral ions present except it lacks the mineral ion under investigation.

Quick Questions

Q1 Describe the position of xylem in a stem.
Q2 Draw a plant cell and label a structure that contains calcium pectate and an organelle that requires a lot of magnesium.

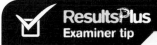

ResultsPlus
Examiner tip

A core practical could be used to ask How Science Works questions. Some examples are:
* Describe or explain a trend shown on a graph of the effect of varying the concentration of a particular mineral ion on the growth of a seedling.
* Calculate the mean and make sure you understand the terms median and mode.
* Give evidence from a table of results which supports a given conclusion and evidence that does not support the conclusion.
* Identify the risks in the practical and how to minimise them.

Thinking Task

Q1 In a mineral ion deficiency study, there are many possible ways to measure the effect on the plant. Suggest *four* different ways to measure the effect.

Starch, cellulose and fibres

Starch and cellulose are both polysaccharides.

Structural feature	Starch	Cellulose
made up of many:	α (alpha) glucose units	β (beta) glucose units
glycosidic bonds	yes	yes
branched molecules	yes (as amylopectin)	never

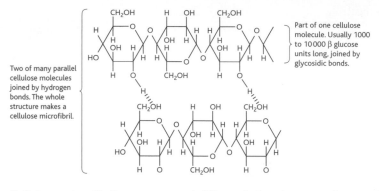

Two of many parallel cellulose molecules joined by hydrogen bonds. The whole structure makes a cellulose microfibril.

Part of one cellulose molecule. Usually 1000 to 10 000 β glucose units long, joined by glycosidic bonds.

The difference between a cellulose molecule and a cellulose microfibril.

Cellulose microfibrils are arranged differently in primary and secondary cellulose cell walls:
- Primary plant cell wall – microfibrils laid down in a criss-cross manner, embedded in a glue of hemicelluloses and pectins. Ligin never present in these call walls.
- Secondary plant cell wall – microfibrils laid down in sheets with each sheet having the cellulose microfibrils running in one direction only. These are embedded in hemicelluloses and pectins. Ligin also present in these cell walls.

Functions of starch, cellulose and cellulose microfibrils and plant fibres in plant cells

- Starch is the energy storage molecule. It is made up of many α-glucose units which can be used in respiration.
- Cellulose is a major component of plant cell walls formed of many β-glucose units.
- Cellulose microfibrils have many hydrogen bonds that together make it strong. In addition, the arrangement of cellulose microfibrils and the glue that binds them give strength and flexibility to plant cell walls whilst allowing them to be fully permeable to water.
- Fibres (sclerenchyma and xylem) give support and some allow the transport of water and mineral ions (due to lignin being strong and waterproof).

Useful properties of cellulose and plant fibres

Humans make good use of the physical properties of cellulose and plant fibres.
1 The properties of cellulose, arranged as microfibrils, produce rope that does not stretch but is flexible and has great strength.
2 Lignified plant fibres are very resistant to chemical and enzyme breakdown. This property makes plant fibres, as wood, a good choice as a building material.

Plant fibres, starch and sustainability

- The use of fossil fuels is not sustainable as they are non-renewable resources.
- Both starch and plant fibres are renewable resources as they come from plants and more can be produced next growing season.
- Starch can be processed into bioplastic to replace oil-based plastics whilst plant fibres can be used for rope.
- Both can be burned to release heat energy.

How to determine the tensile strength of plant fibres

This core practical requires you to find the mass that causes a particular fibre to break. A typical set of equipment is shown in the diagram.

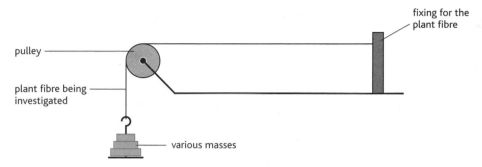

How Science Works as illustrated by some results from this type of practical

Worked Example

Repeat	Mass required to break the plant fibre/ arbitrary units
1	0.206
2	0.203
3	0.207
4	0.205
5	0.209

1 Calculate the mean and give the range of the data.

2 Give one potential risk with this investigation and describe how this risk can be minimised.

3 Describe how a student would use this equipment to discover whether a plant fibre or nylon thread had the higher tensile strength.

Answers

1 The mean is 0.206 and the range is from the lowest force of 0.203 to the highest of 0.209, or it could be written as ± 0.003.

2 Masses falling on foot when the fibre breaks. Place a mat below the masses.

3 State relevant variables that need to be controlled such as threads must have the same diameter. Keep increasing mass until one fibre breaks. The other has the higher tensile strength. Repeat and find the mean.

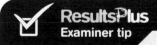

ResultsPlus
Examiner tip

In question 3 of the worked example, you are asked to state which has the higher tensile strength, so an answer that states that the masses should be increased and the threads compared is too general.

Quick Questions

Q1 Give *two* structural similarities and *two* structural differences between the polysaccharides starch and cellulose.

Q2 Suggest why adding smaller masses in the investigation described above produces more accurate results than adding larger masses.

Thinking Task

Q1 Compare the bonds found in starch, cellulose and cellulose microfibrils.

Drugs from plants

A number of plants have antimicrobial properties. This means they are able to combat bacteria and may be of use to humans.

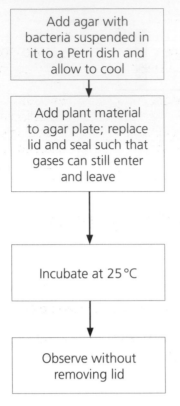

Add agar with bacteria suspended in it to a Petri dish and allow to cool	• The bacteria used must not be harmful to humans • Cooling allows the agar to set firmly.

Add plant material to agar plate; replace lid and seal such that gases can still enter and leave

Ways to add plant material include:
• filter paper soaked in plant extract
• plant extract placed in a hole cut in the agar
• plant material laid on surface.
Seal Petri dish and lid so cannot be opened but air can still get in to stop the development of anaerobic conditions, which encourage harmful bacteria to grow.

Incubate at 25 °C

Incubate at this temp because:
• bacteria grow well
• higher temps, particularly body temp (37 °C), encourages growth of harmful bacteria.

Observe without removing lid

Removal of lid may allow entry/exit of bacteria, including potentially harmful ones.

Investigating the antimicrobial properties of plants

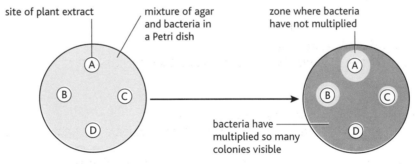

Key to plant extracts
• A to D could be decreasing concentrations of a plant extract
• A to D could be different plant extracts
• D could be a control (without plant extract)

What the Petri dish may look like before and after the investigation.

William Withering's digitalis soup

• Withering (1741–99) a doctor, learned of a remedy for treating a heart condition.
• He isolated the active ingredient from the 20 or so ingredients in the remedy. It was digitalis from the foxglove.
• He trialled different doses on patients to find the most effective treatment.
• He carefully recorded all his findings.

Contemporary drug testing protocols

1 Three-phase testing

If phase 1 testing is passed, then move on to phase 2 testing and only onto phase 3 if phase 2 successful.

Phase 1
Details • a few healthy people used • range of doses **Function** • to check it is safe • that it behaves in the manner predicted from animal tests

Phase 2
Details • c.100–300 patients with the condition used **Function** • to check it is patient safe • to check that it works on the condition as anticipated

Phase 3
Details • c.1000–3000 patients used • normally involves a double blind trial **Function** • to collect as much data as possible including effectiveness compared to placebo or standard treatment and frequency of any side effects

2 Placebo

Name given to tablet/treatment that appears identical in all ways to the drug except that it is chemically inactive.

3 Double blind trial

The procedure is:

- Patients are randomly divided into two groups.
- One group receives drug/treatment and the other group receives the placebo or standard treatment.
- Neither the patients nor those recording any changes in the patients know who has received the drug and who has received the placebo or standard treatment. This reduces the chances of bias.

Modern protocols vs William Withering's digitalis soup

Similarities:

1 Both isolated a possible drug/treatment.

2 Both initially tested on a small number of patients and then a larger group of patients.

Differences:

1 Only modern protocols test on animals before phase 1.

2 Only modern protocols have phase 1 testing where drug is tested on healthy people.

3 Only modern protocols have double blind trials, including using a placebo, undertaken to collect data for statistical analysis.

Quick Questions

Q1 State the phase that uses a large number of patients and a placebo.

Q2 When investigating the effect of different concentrations of garlic extract on bacterial growth, a control was used. State the concentration of garlic in the control.

Thinking Task

Q1 Describe how you would reduce the concentration of $10\,cm^3$ of a plant extract to half the concentration whilst keeping the final volume constant.

Q2 Suggest why large numbers of patients are needed in phase 3 of a modern drug testing protocol.

Topic 4 – Biodiversity checklist

By the end of this topic you should be able to:

Revision spread	Checkpoints	Spec. point	Revised		Practice exam questions	
Drugs from plants	Describe how the uses of plant fibres and starch may contribute to sustainability.	LO6		☐		☐
	Recall the various stages in contemporary drug trial protocols. Recall William Withering's digitalis soup and be able to compare his drug testing process with the contemporary protocols.	LO12		☐		☐
	Describe a practical investigating the antimicrobial properties of plants.	LO1, LO11		☐		☐
Species and biodiversity	Explain the terms *biodiversity* and *endemism*. Describe how biodiversity can be measured.	LO13		☐		☐
	Recall the three domains and molecular phylogeny. Discuss the scientific process used to test new data, as illustrated by molecular phylogeny.	LO16		☐		☐
Natural selection and evolution	Describe the term *niche* and discuss examples of adaptation of organisms to their environment.	LO14		☐		☐
	Describe natural selection and its role in adaptation and evolution.	LO15		☐		☐
Conservation and genetic diversity	Discuss the role of zoos and seedbanks in conserving rare species and maintaining genetic diversity.	LO17		☐		☐
Plant cell structure	Describe the structure of various plant cell organelles and be able to compare a typical plant cell with a typical animal cell.	LO2		☐		☐
Plant stem structure and function	Recall the position and function of xylem vessels and sclerenchyma fibres in a stem.	LO5		☐		☐
	Explain the importance for plants of water, nitrate ions, calcium ions and magnesium ions.	LO9		☐		☐
	Describe a practical on plant mineral ion deficiency.	LO1, LO10		☐		☐
Starch, cellulose and fibres	Recall and compare the structure of starch and cellulose.	LO3		☐		☐
	Understand the structure of cellulose microfibrils and their arrangement within plant cell walls.	LO4		☐		☐
	Identify sclerenchyma fibres and xylem vessels as seen through a light microscope.	LO7		☐		☐
	Describe a practical on testing the tensile strength of plant fibres.	LO1, LO8		☐		☐

ResultsPlus
Build Better Answers

1 The diagram shows two wheat heads. Both were grown in an identical mineral ion solution except that the plant that produced the top wheat head had potassium ions added to the mineral ion solution. The lower wheat head came from a plant grown in a mineral ions solution lacking potassium ions.

(a) Describe the effect of potassium ions on wheat head structure.

 Examiner tip

Potassium ions are not in the specification, but this question is testing your observational skills.

Examiner comments

Make sure your answer is appropriately clear. For example, what does the phrase 'bigger wheat head in the presence of potassium ions' mean in this instance? A better description would give more detail e.g. the wheat grown with potassium ions increased the head length; head width: allowed more seeds to be present; four more seeds present.

(b) The grains in both wheat heads contain starch and cellulose. The cellulose is arranged as cellulose microfibrils. Describe the arrangement of cellulose molecules in a cellulose microfibril and explain how this arrangement gives the microfibril great strength.

 Examiner tip

Look carefully at this question as you are being asked to do two different things. Often the 'describe' command requires the recall of accurate factual knowledge whilst the 'explain' command demands you to give explanatory reasons.

Student answer	Examiner comments
The microfibril is made up of thousands of cellulose molecules all lined up next to each other and joined by hydrogen bonds. This gives great strength because there are thousands of hydrogen bonds.	This question is likely to have several marks attached to it; you will score better if you tackle it in order. The student has sensibly separated the two aspects of the question and tackled each in turn. The description of the arrangement of cellulose molecules into a microfibril is sound but the phrase 'all lined up next to each other' could possibly mean one after the other rather than parallel to each other. The explanation achieves the first point but needs to be clarified and expanded on to gain full marks. ■ A **basic answer** is likely to only focus on the description component and may lack detail, such as only referring to many cellulose molecules rather than supplying a specific range of cellulose molecules. ▲ An **excellent answer** would describe the cellulose arrangement in a precise and clear manner. Further, it would cover most of the 'explain' points such as the additive effect of thousands of weak hydrogen bonds giving great strength between the adjacent cellulose molecules. It might also consider the role of glycosidic bonds within each cellulose molecule.

2 Some drugs have a substance added to help speed up dissolving because the faster they dissolve, the more rapidly they can act.

A student investigated the effect of substance A on the percentage of a drug that dissolved with time. The results are shown in the graph.

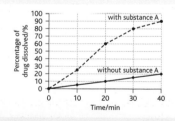

Compare the effect of the presence of substance A on the percentage of drug that dissolved with the absence of substance A, over the 40-minute period.

 Examiner tip

Make sure you compare. 'A has a higher percentage dissolved compared to B throughout the 40–minute' is comparative but writing that 'A has a high percentage dissolved throughout the 40–minute' is not.

Examiner comments

Compare trends such as 'both are increasing throughout the 40 min'. Do not repeat the data such as 'at 10 min, 25% of the drug has dissolved with A present and 5% without A present'.

Try to manipulate the data. A good way to do this is to state the difference at the end of the study. In this case it is that 70% more of the drug has dissolved when substance A is present than when A is absent.

Practice questions

1 The stem of a plant contains many different cells and fibres.

(a) The diagram below shows a section through a stem.

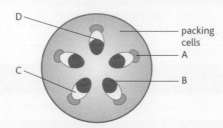

(i) Give the letter of the structure that transports water and helps support the plant. (1)

(ii) Give the letter of the structure that transports mineral ions. (1)

(b) Packing cells are typical plant cells as they have cell walls. Copy the table below which gives some statements relating to plant cell walls. If the statement is correct place a tick (✔) in the box and if the statement is incorrect, place a cross (✘) in the box. (4)

Statement	Statement correct (✔) or incorrect (✘)
hemicellulose molecules present	
calcium is a component of the cellulose molecule	
all plant cells have a primary cell wall	
a tonoplast may be present	

(c) Communication between adjacent plant cells can involve plasmodesmata.

(i) Describe the structure of plasmodesmata. (2)

(ii) State two ways in which pits differs from plasmodesmata. (2)

2 It has been observed that the mean size of a particular bird population has become larger after three consecutive cold winters. Using the theory of natural selection and evolution, suggest how and why the mean size has increased. (4)

3 Mineral ions and water are required by plants for a variety of purposes.

(a) In an investigation to show that nitrate ions are needed for plant growth, a student grew 10 wheat seedlings in a mineral ion solution containing nitrate ions. He also grew another 10 wheat seedlings in a mineral ion solution lacking nitrate ions. All other variables were kept constant.

(i) Suggest the role of the mineral ion solution lacking nitrate ions. (1)

(ii) Suggest why the student made sure that the 20 wheat seedlings had come from seeds from one wheat plant. (1)

(iii) The investigation was run for 14 days and the mean increase in seedling length was 30 mm. Suggest how the student calculated this figure. (2)

(iv) The seedlings in the mineral ion solution lacking nitrate ions increased by 15 mm over the same 14 days. Give an appropriate conclusion for the effect of nitrate ions on wheat seedling growth. (2)

(b) The passage below refers to three mineral ions in plants. Copy and complete the passage by filling in the missing word or words.

_____ ions supply the element _____ which is a component of amino acids. The amino acids are the building blocks of _____. Calcium is found in the _____ lamella whilst _____ is found in chlorophyll. (5)

(c) Describe the functions of water in a plant. (4)

4 When a potential new drug is produced it has to go through rigorous drug testing protocols before it can become available for use.

(a) Copy the table below which describes some features of the three-phased drug testing protocol. Give the correct number of each phase being described. (3)

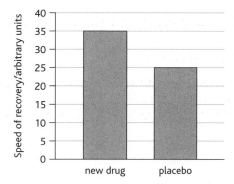

Description	Phase
A few patients are involved.	
A few healthy volunteers are involved.	
A double blind trail is undertaken.	

(b) The graph below shows the speed of recovery for patients on a new drug and on a placebo.

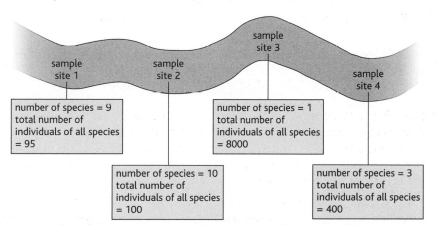

Give the speed of recovery that is due to the medical effect of the drug. Give a reason for your answer. (2)

5 Biodiversity is a measure of biological variation and includes species richness and genetic diversity.

The diagram shows a stream that has been sampled at four different sites. The number of species and the total number of individuals of all species was recorded at each sample site.

sample site 1

sample site 2

sample site 3

sample site 4

number of species = 9
total number of individuals of all species = 95

number of species = 10
total number of individuals of all species = 100

number of species = 1
total number of individuals of all species = 8000

number of species = 3
total number of individuals of all species = 400

(a) Suggest which site has the highest species richness. Give a reason for your answer. (2)

(b) Suggest why taking *one* sample at each site may reduce the reliability of the data. (1)

(c) State the sample site that has the highest population density of organisms. (1)

Unit 2 Specimen paper

1 Both plant and animal cells that make glycoprotein require rER and Golgi apparatus. Mitochondria can make some protein.

(a) Copy and complete the passage by filling in the missing word or words.
_____ _____ are joined together to make protein on the surface of rough _____ reticulum (rER). The protein may enter and become packaged into small spheres called _____ . These move to the Golgi apparatus. In the Golgi apparatus the protein is _____ into glycoprotein. (4)

(b) The electron microscope has revealed many organelles such as mitochondria. The diagram below shows a partially labelled mitochondrion.

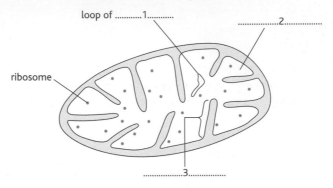

loop of ....1...........

.................2.................

ribosome

..................3.................

(i) Copy and complete the labels by replacing the numbers with the correct words. (3)

(ii) Mitochondria are said to have an envelope. What is meant by the term *envelope*. (2)

(c) Ribosomes are also found on the rER. Suggest the difference between the ribosome found in the mitochondrion and a ribosome found on the rER surface. (1)

Total 10 marks

2 For fertilisation to occur, gametes need to be made through meiosis.

(a) During meiosis, a variety of events occur. For each of the following, write down the letter that identifies the correct answer.

(i) Different gene combinations are formed due to:
A centromeres forming.
B centrioles breaking.
C crossing over.
D telophase. (1)

(ii) The number of chromosomes is:
A doubled.
B halved.
C quadrupled.
D quartered. (1)

(b) As part of the plant fertilisation process, a pollen grain produces a pollen tube.
(i) Describe the route the pollen tube follows from the pollen grain to the egg cell. (2)

An investigation was undertaken to study the effect of varying the storage time of pollen on pollen tube growth from one plant species. One set of pollen was allowed to germinate and produce pollen tubes immediately after collection whilst the other set was stored at 4°C for 24 hours before being allowed to germinate and produce pollen tubes.

The results of the investigation are shown in the table below.

Time/hours	Mean pollen tube length/mm	
	Immediately after collection	After storage
0	0.0	0.0
6	0.6	0.4
12	0.8	0.6
18	0.9	0.7
24	1.0	0.7

(ii) Compare the mean pollen tube length when germinated immediately after collection with the mean pollen tube length after storage, over the 24-hour period. (3)

Total 7 marks

3 Plants contain several polysaccharides.
   (a) The table below describes two polysaccharides found in plants. Copy and complete the table by filling in the missing word or words in the empty boxes. (3)

Polysaccharide	Chemical composition	Shape
	made up of many α glucose	
cellulose		linear molecule

   (b) In cell walls, cellulose is organised into cellulose microfibrils.
      (i) Describe the structure of a cellulose microfibril. (2)
      (ii) The diagrams below show four possible arrangements of cellulose microfibrils in a primary cell wall. Write the letter of the diagram that shows the correct arrangement. (1)

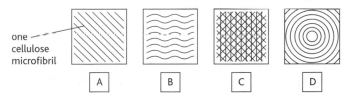

   (c) The arrangement of cellulose microfibrils helps to give plant cells their strength. Some plant fibres are further strengthened by the addition of lignin.

   The graph below shows the effect of increasing the diameter of the plant fibre on the mean mass needed to break the fibre. Ten readings were taken for each diameter and error bars have been included.

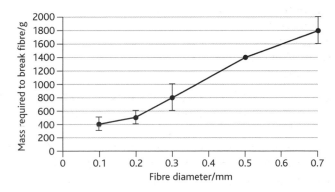

(i) Describe the effect of increasing the diameter of the fibre on the mean mass needed to break the fibre. (2)

(ii) Explain why the mean value at 0.2 mm may be regarded as being more reliable than the mean value at 0.3 mm. (2)

(iii) At the fourth data point, there appears to be no error bar, despite 10 readings having being taken. Suggest a reason for this. (1)

(iv) A second study showed that a man-made fibre required a slightly higher mass to break the fibre at all diameters. However, it was not used as it did not contribute to sustainability. Explain why the man-made fibre did not contribute to sustainability. (2)

Total 13 marks

4 Some bacteria have become resistant, through natural selection, to antibiotics.

(a) Bacteria make up one of the three domains of life. Name the other two domains. (2)

(b) The diagram below shows a typical bacterium.

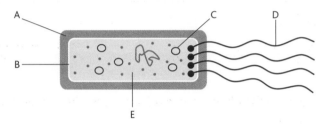

Copy and complete the table by adding in the letter, from the diagram, that correctly identifies the statement. The first one has been done for you. (3)

Statement	Letter
site where biochemical reactions take place	E
contains genetic material, including genes for antibiotic resistance	
structure made of peptidoglycan that is only found in bacteria	
membrane that regulates what enters and leaves the bacterium	
structure used for movement	

(c) Describe how natural selection has enabled some bacteria to evolve resistance to antibiotics. (5)

(d) When testing drugs, such as a new form of antibiotic, several phases have to be completed.

(i) Suggest why it is sensible to test the drug on healthy volunteers (phase 1) before testing it on patients (phase 2). (1)

(ii) Suggest why a placebo is used as part of phase 3 of a drug trial protocol. (2)

Total 13 marks

5 Biodiversity is an important concept in conservation.
   (a) State two techniques a zoo could use to try to maintain the genetic diversity of a species. (2)

   (b) Suggest one example of the educational role of zoos. (1)

   (c) Seed banks store large numbers of seeds over long periods of time. The graph shows information on percentage success of seeds germinating, from one plant, after the seeds had been stored for different lengths of time. The seeds were kept at room temperature.

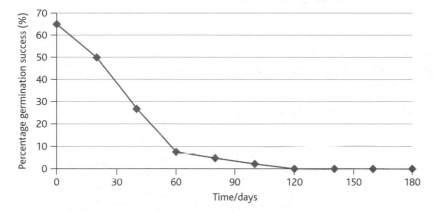

   (i) Describe the effect of increasing storage time on percentage germination success. (3)

   (ii) Suggest how seedbanks may overcome this change in percentage germination success. (2)

   Total 8 marks

6 Embryonic stem cells include pluripotent and totipotent stem cells. The embryonic stem cells used in research can come from spare embryos.
   (a) Explain what is meant by the term *spare embryos*. (2)

   (b) Some people believe it is a good idea to use spare embryos for research and some people believe it is a bad idea to use spare embryos for research.
   (i) Give *two* reasons why some people believe it is a good idea. (2)
   (ii) Give *two* reasons why some people believe it is a bad idea. (2)

   Total 6 marks

7 It has been suggested that the ratio of males to females in a small wild population can affect genetic diversity.

   (a) Explain what is meant by the term genetic diversity. (2)

   A mathematical model was used for a population of 10 unrelated individuals. The theoretical genetic diversity was calculated for four different ratios of male to female. The results are shown in the table.

Number of individuals in the population		Theoretical genetic diversity /arbitrary units
Males	Females	
2	8	4.0
3	7	7.0
4	6	9.0
5	5	10.0

   (b) Describe the effect of the ratio of males to females on the theoretical genetic diversity. (2)

   (c) Using your own knowledge, explain the result for the ratio of two males and eight females. (2)

   Total 6 marks

Research skills: Visit or issue report

What do 'real' biologists do? They identify a problem, try to come up with a solution using source material, consider the implications, benefits and risks of this solution, look at alternatives and attempt to communicate their ideas effectively. Your Unit 3 task is to find out how biological scientists research a real-life problem. You have to report on a visit made or a topical issue researched, showing knowledge and understanding of preliminary research on the problem, evaluation of source material and effective presentation of your ideas.

- There are 40 marks available – 20% of the total AS marks.
- The report should be 1500–2000 words.
- The report must be word processed.

Getting organised

If you choose a visit – Your teacher should carry out a preliminary visit if possible, to make sure an obvious problem is being solved. You could do preliminary work yourself by looking at the website of the organisation and comparing it with those of similar ones.

Write a list of questions before the visit to help focus your mind on the research skills involved. This will ensure that your report is original and more interesting. A 'talk' is sometimes part of the visit. Valuable as this is, you will all have the same talk and the notes you take will produce reports that are all very similar. That is why your own questions are so important and you will be better prepared.

If you choose an issue – Make sure this is a real-life problem that biologists are trying to solve. Encourage your teacher to have a weekly discussion of 'biology news' items to look at. These could prove useful for an issue report.

So where do the marks go?

1.1 Identify and describe the question or problem (4)

It must be obvious what problems biologists are trying to solve for full marks.

> '**Problem of human–elephant conflict:** The initial cause of the HEC is agreed by many sources such as www.peopleandwildlife.org.uk to be the increase in human population, resulting in the increase of land use for agriculture and consequent decrease of elephant habitations. The elephants are then forced to live in much greater proximities to one another and according to many sources [5] this can result in shortages of food eventually leading to animals leaving their current regions and therefore raiding crops and damaging farms.'

This extract from an issue report on elephant conservation scored 4 marks because it also went on to describe groups of six elephants led by a mature bull and how the dry climate in some areas can exacerbate the problem.

1.2ab Methods used to try and solve the problem (4)

You must discuss exactly what biologists are trying to do to solve the problem (2) and give some data to support the discussion (2).

> **A Possible Solution – Antipsychotic Medications:** Antipsychotic drugs are the most effective treatment to alleviate psychotic symptoms of schizophrenia.[2-16] The drugs work to decrease the chemical imbalance in the brain that is causing symptoms but would also induce side effects..........
> The medications are divided into two types, traditional and new, based on their specific neurotransmitter receptor affinity and activity. Each drug differs in how it affects other brain chemicals. Individuals thus respond differently and sometimes several medications must be tried before the right one is found.'

This extract on schizophrenia gained the maximum 2 for 1.2a and 1.2b because the report then went on to describe several traditional and new forms of drug therapy with a discussion of how effective they are, giving some data and examples to support the discussion.

1.3a Valid or reliable data (2)

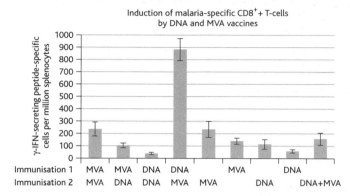

Induction of malaria-specific CD8+ + T-cells by DNA and MVA vaccines

The report must include some graphs, tables or diagrams in the discussion on whether the methods are appropriate or effective. This extract on malaria shows just that:

> *'The graphs show how the 'prime boost' technique induces a greater T-cell response than simply a single immunisation. The graphs also show that the sequence of vaccination is critical to optimising the results.[18]'*

ResultsPlus
Examiner tip

Make sure that any graphs, tables or diagrams used here are actually discussed by you and that you show why they are relevant.

1.3b Is the solution appropriate? (2)

You need to discuss just how appropriate the solution is.

This issue report on new drug treatments for schizophrenia easily gains the 2 marks because in stating '*they are considered more effective and favourable over the traditionals*', the student has shown why they are now a much more appropriate and effective treatment.

> *'The new antipsychotic drugs, which include clozapine, risperidone and olanzapine, etc. block dopamine receptors more selectively than traditional antipsychotics. Some also block or partially block serotonin receptors (particularly $5HT_{2A, C}$ and $5HT_{1A}$ receptors). [19]This decreases the likelihood of extrapyramidal (motor) adverse effects, such as Parkinsonian-like symptoms, acute dystonic reactions and akathisia. [8, 17, 20, 26] Therefore, they are considered more effective and favourable over the traditionals. Also, the new medications tend to alleviate positive symptoms and may lessen negative symptoms to a greater extent. [12]'*

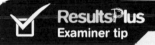

ResultsPlus
Examiner tip

This must be a thorough discussion with your own ideas to show why the solution in 1.2 is appropriate – does it actually solve the problem? Use data or examples to show this.

2.1 Implications of solution to problem (4)

You need to identify and explain at least two implications. They must be linked to the applied biology solution, not the original problem. This extract from an issue report on the problem of AIDS in Africa scores the maximum of 4 marks.

> *'**Ethical, Economic & Social Issues** – Symptoms of AIDS may not develop until long after infection, and a sufferer might not know they have the virus, but can still transmit it to others. In my view, governments should attempt more vigorous screening, preventing thousands from unknowingly transmitting HIV through intercourse and pregnancy. This is undoubtedly one of the most prominent ways in which HIV is spread. However, HIV test results are never 100% accurate and can incur prejudice. When applying for insurance, people are asked if they have had an HIV test or counselling. Although life-insurance companies are now under pressure to replace such questions by asking only if they have had a positive test, this offers no reassurance to those with the possibility of a positive result. In some areas, being known as an HIV sufferer is too great a risk. Under-developed countries have the highest HIV rates but insufficient money for adequate medical resources to implement preventative methods. Only 8% of HIV-positive women have access to treatment, counselling and testing.'*

Notice how the extract achieves:

- 1 mark for identifying the social implication of preventing AIDS transmission through screening and another mark for explaining how the screening prevents unknowing transmission.
- 1 mark for identifying the ethical implication of false results or of insurance companies asking for screening test results and another mark for explaining how this can cause prejudice against HIV sufferers.

2.2ab Benefits and risks of solution (4)

The issues identified in 2.1 need to be explored more fully. There are 2 marks for benefits and 2 for the risks. Two examples are the issue report on elephants raiding and trampling crops in Africa and the visit report on making cheese for vegetarians.

> *'The advantage of using traditional methods like noises or chilli peppers for repelling elephants away from farming fields is that they are cheap and can be used by farmers easily. Unfortunately, the problem with such methods is that elephants get used to simple tactics and eventually learn to ignore the noises as they realise they will not be harmed.'*

> *'**GM Rennet** – for production of the enzyme, a non-pathogenic strain of E. coli bacteria, as well as yeast cells has been engineered. In fact, 90% of all vegetarian cheese sold in Britain now uses GM rennet. [14] This is controversial because many people feel genetic modification also has ethical implications. In fact, there are further ethical implications for vegetarians themselves because they feel that the use of an animal gene is unacceptable.'*

2.3 Alternative solutions (4)

You now need to consider two alternatives. For example, these are two alternatives to using antiviral drugs to prevent transmission of HIV to the fetus.

> *'**Caesarean section** – A caesarean section delivers the baby through an incision in the mother's lower abdominal wall and uterus. In HIV-positive mothers it is done to protect the baby from direct contact with her bodily fluids, which contain the HIV virus.*
>
> ***Bottle-feeding** – The World Health Organisation recommends avoidance of all breastfeeding when replacement feeding is affordable, feasible, acceptable, sustainable and safe. HIV-infected T-helper cells are within breast milk and may otherwise be passed on to the baby.'*

3.1 Using source material (4)

It must be obvious in your discussion that you have read and used information from your sources (1). You must refer to at least three sources (1) with at least one web-based (1) and one non web-based (1). The best reports will include references to at least five or six sources. Try to use sources other than your textbook because it shows genuine original research on your part.

Plagiarism: This is serious and the penalties could be severe. If you use source material in your report make sure that:

- it is in parentheses (brackets) so it is obviously not your work but you still want to use it
- you acknowledge it and refer to it in the bibliography.

3.2 Bibliography (4)

This is a list of all the sources you used (1) with full (1) details given, including the publisher if it is a book. You must show in your report where you have used each source (2 marks for all, 1 mark for some), e.g. number the sources and give the number when you use information from that source. Here are a couple of examples but also look at the Parkinson's example to see how you could use reference numbers in the text.

> 1. 'Cognitive behaviour therapy for schizophrenia'
>   http://www.cochrane.org/reviews/en/ab000524.html
> 2. Makrides, M., Neumann, M., Byard, R., Simmer, K., & Gibson, R. (1994): Fatty acid composition of brain, retina, and erythrocytes in breast and formula-fed infants. *Amer. J. Clin. Nutr.*, **60**; 189–194.
> 3. Taylor, B. (1999): *Nature Watch Elephants*, Lorenz Books.

3.3 Validity of sources (4)

You must choose two of your sources and then comment on their reliability or validity. Can you trust the information? Is it true? The best way to do this is to cross-check with other sources using a simple Google search and see if the evidence agrees or not. Then base your opinions on this. Here is one example.

> *'The information obtained from Lincolnshire Poacher [1] was obtained through demonstration by the head cheese maker. I can thus put a degree of trust in it. However, I have confirmed the validity of the data. For example, where Lincolnshire Poacher claims artificial rennet creates bitter cheese, further research [18] confirms this.'.*

4 Communication (4)

These marks are for good spelling, punctuation and grammar and use of biological terms showing that they were understood. Try writing for a particular audience because this will encourage you to make the layout informative and readable with graphs, tables and diagrams to illustrate your points.

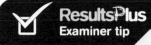

ResultsPlus
Examiner tip

Making a comment on the reliability gets you the first mark but the evidence from checking gets the second mark for each of the sources. Look at the last sentence in the extract.

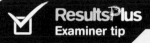

ResultsPlus
Examiner tip

A 'Word' document is fine but your format could be a film, magazine or newspaper article, PowerPoint presentation, poster, or even a web page – as long as you follow the assessment criteria.

ResultsPlus
Examiner tip

When you see pairs of marks, one is for a brief discussion but the second is for a full explanation with your *own* ideas

Answers to in-text questions

Unit 1: Topic 1

Transport and circulation pp.8–9

QQ

1 They make it a liquid by holding molecules together. They hold molecules together giving it cohesion. They allow it cling to other substances, adhesion.
2 The oxygen carries the +, the hydrogen the –.
3 Highest – yeast, *Daphnia*, goldfish, golden retriever, elephant, killer whale – lowest.

TT

1

$$B = \frac{6 \times 4}{2^3} = \frac{24}{8} = 3$$

$$C = \frac{6 \times 16}{4^3} = \frac{96}{64} = 1.5$$

As organisms become larger, their SA/V ratio becomes smaller and diffusion becomes less able to supply their needs.

Cardiac cycle and heart rate pp.10–11

QQ

1 a right ventricle
  b right atrium
  c left ventricle
2 Three from: abundant, easy to get, parthenogenetic, simple nervous system, does not suffer apparent stress, transparent so procedure can be non-invasive.

TT

1 Blood drains into the right atrium from the body along the vena cava ⟶ raising of the blood pressure in the right atrium forces the right atrioventricular valve open ⟶ contraction of the right atrial muscle (atrial systole) forces more blood through the valve ⟶ as soon as atrial systole is over the right atrioventricular muscle starts to contract (right ventricular systole) ⟶ this forces the right atrioventricular valve closed and opens the valve in the mouth of the pulmonary artery (semilunar valve) ⟶ blood then leaves the right ventricle along the pulmonary artery.

Cardiovascular disease (CVD) pp.12–13

QQ

1 fibrin, calcium, vitamin K, thrombin, thromboplastin
2 platelets

TT

1 A: platelets stick to damaged wall; B: thromboplastin released; C: calcium ions *or* vitamin K; D: calcium ions *or* vitamin K; E: prothrombin; F: thrombin; G: fibrinogen; H: fibrin

Structure and function of carbohydrates p.14

QQ

1 lactose is glucose and galactose
  sucrose is glucose and fructose
  maltose is two glucose

TT

1 glucose – has reactive group so good for respiratory substrate
  sucrose – reactive groups joined to each other so unreactive, so good for transport function in plants
  amylose and amylopectin – both have compact structure so can fit lots of glucose into small space, good for storage, large so do not dissolve, no osmotic effect, cannot pass through membranes so good again for storage, held together by glycosidic bonds so easily hydrolysed, amylopectin hydrolyses more rapidly due to branched structure.

Structure and function of lipids p.15

1 In both cases water is lost and it is therefore a condensation reaction.
2 4: glycerol and three fatty acids

TT

1 three; one to be involved in breaking each ester bond between a fatty acid and glycerol

Risk factors for cardiovascular disease pp.16–17

QQ

1 Chemicals in cigarette smoke cause arteries to constrict thus raising blood pressure; they can physically damage the lining of arteries leading to atherosclerosis.
2 Oestrogen seems to afford protection, its secretion stops at menopause.

TT

1 Lower HDL, higher CHD risk (or vice versa); increasing LDL increases this risk further; even patients with low LDL (bad cholesterol) have significantly increased risk if they also have low HDL.

Population studies on risk factors pp.18–19

QQ

1

Case-control studies	Cohort study
looking at past history to explain present observations, retrospective	following study groups into the future, prospective
control group selected for match with 'experimental' group	control group selected as those who do not end up with the condition

2 large sample size, long term, ethical, representative sample, valid measurement techniques used, methods used reliable (e.g. standardised procedures or operatives the same)

TT

1 a Because on all risk factors given they have higher consumption than UK people.

b Possible as they drink wine more regularly which has antioxidants, may have high HDL in fat. May eat more of 'good' foods, e.g. fish. Maybe heart disease under reported. Maybe fat is not important after all.

Using scientific knowledge to reduce risk pp.20–21

QQ

1 Personal experiences, not weighing one risk against another, peer pressure, remoteness, fatalistic.

2 Vitamin C is an *antioxidant* which gives electrons to other substances; this makes it a *reducing* agent. DCPIP is an *oxidising* agent which can gain *electrons* from *reducing* agents.

TT

1

Juice tested	Average volume of juice required to decolourise DCPIP /cm⁻³	mg vitamin C cm⁻³ juice
grapefruit juice	1.61	3.72
pineapple juice	11.56	0.52
orange juice	2.12	2.83
orange drink	1.45	4.14
fresh lemon juice	1.73	3.47
bottled lemon juice	24.0	0.25

Unit 1: Topic 2

Structure of amino acids and proteins pp.26–27

QQ

1 any two monosaccharides or disaccharides; two amino acids; glycerol and a fatty acid

2 peptide bond

TT

1 No, because all have NH_2 at one end and COOH at the other, and it is between these that the peptide bond is formed.

Enzyme action and rates of reaction pp.28–29

QQ

1 The rate should express the amount of substrate used or product produced in unit time, it is proportional to 1/time.

2 enzyme concentration, temperature, pH, volume of solutions

TT

1

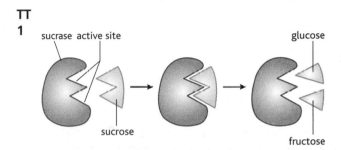

Cell membranes pp.30–31

QQ

1 protein, lipid/phospholipid, carbohydrate, cholesterol, glycolipids

2 4; 2 fatty acids, glycerol, phosphate

TT

1 Some suggestions: receptors for signal molecules (e.g. hormones), electron carriers, cell recognition (mainly as glycoproteins), membrane bound enzymes

Transport across cell membranes pp.32–33

QQ

1 active transport

2 osmosis, exocytosis, endocytosis, diffusion and facilitated diffusion

TT

1

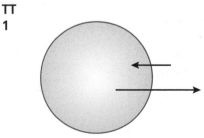

Structure and role of DNA and RNA pp.34–35

QQ

1 a adenine (must be name not just A)

b uracil (not just U)

2 phosphate; deoxyribose (half mark for sugar); base

TT

1

DNA	RNA
double stranded	single stranded
sugar is deoxyribose	sugar is ribose
bases A, C, T, G	bases A, C, U, G

DNA replication and protein synthesis pp.36–37

QQ

1 Complementary strand
mRNA UACGGUAUGCCAACCUUC
ATGCCATACGGTTGGAAG

2 Transcription – the making of a complementary copy of the portion of the DNA which constitutes a gene, mRNA synthesis. **Translation** – the putting together of a chain of amino acids in a sequence dictated by the order of bases in mRNA. **Template strand** – the strand in double-stranded DNA on which the mRNA copy is formed in transcription. **Sense strand** – the strand which is complementary to the template strand.

TT

1 You cannot say without knowing the identity of the complementary base on the other strand, the preceding part of the sequence does not determine the next part.

2

DNA replication	Transcription
DNA polymerase catalyses	RNA polymerase catalyses
both DNA strands involved	only one DNA strand involved (the template strand)
A pairs with T during the process	A pairs with U during the process
whole DNA molecule replicated	only part of the length of the DNA copied (the gene)

Gene, mutation and cystic fibrosis pp.38–39

QQ

1

Tube blocked	Problem
bronchioles	less fresh air gets to alveoli – leads to shortness of breath
pancreatic duct	digestive enzymes don't get into small intestine – leads to inefficient food digestion
cervix	sperm cannot enter uterus – leads to no conception
vas deferens	sperm cannot move to urethra – leads to infertility

2 It does not allow chloride ions out of the cell, and it does not inhibit the sodium channel. Both of these mean that the osmotic gradient is into the cell rather than out so water is withdrawn from the mucus outside the cell, making it too sticky.

TT

1 a

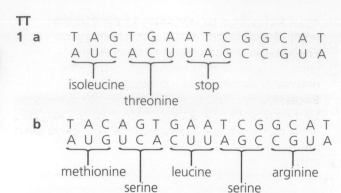

2

Effect of CF sufferer	Origin of effect
breathing problems	Alveoli wall lined with thick mucus making diffusion path too long, bronchioles blocked with sticky mucus stopping entry of fresh air into alveoli.
lung infections	Sticky mucus harbours bacteria.
poor weight gain	Food digestion inefficient because pancreatic duct blocked with sticky mucus preventing digestive enzymes getting into small intestine.
reduction in likelihood of pregnancy in women sufferers	Sticky mucus plug develops in cervix preventing sperm entry.
inefficient movement of sperm from testes to penis, less sperm in ejaculate than normal	Vas deferens blocked with sticky mucus; vas deferens sometimes absent.

Genetic inheritance pp.40–41

QQ

1 Allele: different versions of the same gene (found at the same locus but in homologous chromosomes or in different individuals) that may produce different phenotypes.
Dominant: an allele that masks the presence of a recessive allele in the phenotype. Dominant alleles for a trait are usually expressed if an individual is homozygous dominant or heterozygous.
Gene: that portion of the genome that carries the information for a single protein. (In cases of proteins with multiple subunits, there may be a gene for each.)
Genotype: the genetic makeup of an individual. Genotype can refer to an organism's entire genetic makeup or the alleles at a particular locus.
Heterozygote: carrying two different alleles of a gene.
Homozygote: having the same allele at the same locus on both members of a pair of homologous chromosomes. Homozygous also refers to a genotype consisting of two identical alleles of a gene for a particular trait. An individual may be homozygous dominant (**AA**) or homozygous recessive (**aa**).

Individuals who are homozygous for a trait are referred to as homozygotes. See heterozygous.
Phenotype: the observable or detectable characteristics of an individual organism – the detectable expression of a genotype.
Recessive: an allele that is masked in the phenotype by the presence of a dominant allele. Recessive alleles are expressed in the phenotype when the genotype is homozygous recessive.

2 0%

TT

1 For CF choose **N** and **n**; albinism: **N** and **n**
Thalassaemia: **N** and **n**
Peas: **R** and **r**

Gene therapy and genetic screening pp.42–43

QQ

1 *In vitro* fertilisation, IVF; pre-implantation genetic diagnosis, PIGD; chorionic villus sampling, CVS

2 viruses, liposomes

TT

1 Insurance problems might arise if you are found to be carrying something; they will not necessarily understand the biology of the situation and think you are ill or likely to be.
Difficulty in getting a job.
False possitives and false negatives.

2 There are no right answers to questions like this, what an examiner would be looking for are statements backed up by scientific thinking, not unfounded opinions.
Justify: baby will be very ill, likely to die quite young, will suffer.
Decide not to: religious reasons you may both have, child can have a good life while it is alive, much that can now be done to reduce symptoms.

Unit 2: Topic 3

Cells and organelles pp.52–54

QQ

1 As the human liver cell is a eukaryotic animal cell, answers would be that only the liver cell has a nucleus surrounded by membranes; organelles in the cytoplasm surrounded by one or two membranes; DNA as separate strands; no cell wall present; no slime capsule ever present; (only three answers needed to gain full marks).
Of course, you could have tackled it the other way round and told the examiner the structural features that only prokaryotes have.

2 Double membrane/envelope enclosing the structure.

3 rER is covered in ribosomes, appear as irregular structures; Golgi shows as a stack of cisternae without ribosomes

4 Any four from: nucleus, rER, sER, Golgi, mitochondria, lysosome (vacuole also possible)

TT

1 Though the cell has no nucleus, it still has cytoplasm which contains all the membrane-bound organelles.

2 Three or more cisternae drawn in parallel; getting smaller from one cisterna to the next; cisterna curved; vesicles clearly associated with Golgi body.

3 Protein enters rER cavity ⟶ Vesicles with protein inside form and bud off rER ⟶ Protein moves to Golgi in vesicles ⟶ Protein enters Golgi and is modified ⟶ Modified protein leaves Golgi in vesicles ⟶ modified protein released from cell.

Cellular organisation p.55

QQ

1 A system is made up of several organs, working together, for a specific function.

2 Organs are made up of many cell types; organised into tissues.

TT

1

Feature	Tissue	Organ
made up of cells	✓	✓
working together	✓	✓
for a function	✓	✓
can be made up of more than one tissue	✗	✓

Cell cycle and mitosis pp.56–57

QQ

1 Chromosomes line up at the centre; joined to spindle fibres; at centromere.

2 A newly formed cell grows to maturity ⟶ Cell undergoes its normal metabolic function ⟶ Cell undergoes mitosis ⟶ Cell undergoes cytokinesis to become two cells.

3 Stain – to make the chromosomes more visible. Acid – to soften the tissue.

4 Interphase, largest segment of pie chart.

TT

1 To attach to chromosomes; so can pull them apart; during anaphase; so each daughter cell has the correct number of chromosomes.

Production of gametes p.58

QQ

1

Structure present	Sperm cell	Egg cell
cell surface membrane	✓	✓
haploid nucleus	✓	✓
cytoplasm	✓	✓
mitochondria	✓	✓

TT

1 Any of the following combinations: small black +medium light on left/small light + medium black on right; small light + medium black on left/ small black +medium light on right; small light + medium light on left/ small black +medium black on right.

Fertilisation p.59

QQ

1 Receives digestive enzymes from sperm; and this area becomes digested/hydrolysed; so a channel forms in the jelly layer / zona pellucida; hardens/thickens when sperm enters egg; due to substance/named substance released from egg.

TT

1 Six; because six ovules; and each ovule has one embryo sac; with one egg cell in it.

Stem cells and cell specialisation pp.60–61

QQ

1 Both are undifferentiated; can give rise to more of their own cell type; can give rise to cells that can become specialised.

2 mRNA enters ribosomes; which convert the mRNA code into protein. This protein modifies the cell to make it specialised.

3 One explant may die/eq but for three this is much less likely.

Stem cells and medical therapies pp.62–63

QQ

1 Possible (route to) infection. Idea of increased chance of cancer/tumour cells developing.

2 More of them; in a larger structure/blastocyst.

TT

1

Question	People working in stem cell research	Everyone else
Is the stem cell procedure technically possible?	✓	
Is the stem cell procedure ethically acceptable?	✓	✓
Should the maximum age for embryos used in research be increased (in the future)?	✓	✓

Variation in phenotype pp.64–65

QQ

1 a Polygenic inheritance. Environmental effect such as varying amounts of nutrients the baby gets whilst inside mother.

b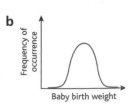

2 (non-optimal) temperature, sunlight, mineral ions, soil type, wind speed, water level.

TT

1 a having type 1 diabetes

b Identical twins have same genotype/genetically identical; and one twin has a much higher/eq likelihood of having the condition if other twin has it, compared to non-identical twins.

c Does not matter whether you are an identical twin or non-identical twin, you will not have the condition if not exposed to environmental trigger.

Unit 2: Topic 4

Species and biodiversity pp.70–71

QQ

1 Quadrat sited randomly ⟶ All species counted ⟶ Process repeated ⟶ Work out mean number of species per unit area

TT

1 Alleles rare in the population. Not sampled by chance.

Natural selection and evolution pp.72–73

QQ

1

Description of adaptation	Anatomical	Behavioural	Physiological
bee orchid flowers look similar to insects	✓		
a peahen is said to select her peacock mate on the size and shape of his tail		✓	

2 It is the (selective) pressure that makes it possible for certain alleles/genetic variants to be more successful in survival and reproduction than others.

TT

1

Description	Environmental change
a small UK mammal has extended its range north over the last 50 years	warmer/global warming
tooth shape has become different in foxes that live in a city compared with foxes that live in the countryside	diet change/eq

Conservation and genetic diversity pp.74–75

QQ

1 This exposes the animal to new individuals; who, therefore, are likely to be genetically different; (and) who may act as mates.
2 Enzymes are slowed down; decay/eq occurs more slowly; germination is inhibited.

Plant cell structure pp.76–77

QQ

1 Secondary cell wall absent in both. Both lined with membrane.

2

Feature present	Chloroplast	Amyloplast
double membrane	✓	✓
starch	✓	✓
DNA	✓	✗
ribosomes	✓	✗
internal membrane	✓	✗

TT

1 Both have double membrane present; both have a loop of DNA.

Plant stem structure and function pp.78–79

QQ

1 Innermost side/eq; of a vascular bundle.
2 Drawing of cell with label to middle lamella (for calcium pectate), plus label to chloroplast (for magnesium).

TT

1 Change in stem/shoot/leaf/root length. Change mass of leaves/stems/roots. Change in number of leaves. Change in leaf area. Change in colour of leaves.

Starch, cellulose and fibres pp.80–81

QQ

1 **Similarities**: Both made of glucose units; joined by glycosidic bonds.
Differences: Only starch has α glucose/only cellulose has β glucose; only starch can have branches.
2 Idea that the mass required to break the fibre is somewhere between the last mass added before it breaks and the one that caused it to break. Idea that using smaller masses means there is a smaller range between the true breaking mass and the recorded breaking mass (thereby increasing accuracy).

TT

1 All have glycosidic bonds; but starch does not have hydrogen bonds. Reward idea of further detail about the glycosidic bonds such as reference to α in starch and β in cellulose.

Drugs from plants pp.82–83

QQ

1 phase 3
2 zero

TT

1 $5\,cm^3$ of (distilled) water and $5\,cm^3$ of plant extract.
2 To collect as much statistical data as possible. Looking for rare side effects.

Answers to practice questions

Unit 1 Topic 1

1 (a) (i) 4

Feature	Glucose	Glycogen	Maltose	Starch
1–4 glycosidic bonds present	✔	✔	✔	✔
1–6 glycosidic bonds present		✔		✔
made up of many monomers		✔		✔

    (ii) lactose   1
    (iii) two glucose correctly drawn;
         water   2
    (iv) hydrolysis   1

(b) large molecule/made up of many monomers allows storage of large amounts of energy; compact therefore large amounts can be stored in a cell; insoluble therefore does not have an osmotic effect/eq;inert so not affected by other reactions in cells; large molecule therefore cannot pass out of cells; can be hydrolysed to release large amounts of energy/glucose when required;5

Total 13 marks

2 (a) 1 during first 5 years/initially there is little change in death rates;
    2 from late 70s/early 80s in most countries there is a lowering of death rates;
    3 Poland is an exception when death rates have increased;
    4 any quantitative manipulation of data, e.g. death rate in Finland has halved over period; 3

  (b) (i) Finland has high rates of CHD and high BP; UK has high rates of CHD and high BP; 2
    (ii) there is conflicting evidence; Italy has high BP but low death rates; 2

  (c) damage to artery walls;
     ref. to blood clot;
     blood clot can block arteries;
     ref. to coronary arteries;
     lack of (oxygenated) blood flow to heart muscle; 3

Total 10 marks

3 (a) (i) 32.9/33 kg m^{-2} 2
    (ii) 1 Increase level of exercise;
      2 (which) increases metabolic rate/uses more energy/eq
      3 (which) increases weight loss/eq
      4 ref. to change in diet;
      5 decrease saturated fat intake
      6 (therefore) reducing blood cholesterol level

      7 low carbohydrate diet/reduce alcohol intake/reduce sugars in diet;
      8 to restrict calorie intake;
      9 to reduce risk of type 2 diabetes;
     10 lowers blood pressure;
     11 reduces risk of cardiovascular disease; 5

  (b) (i) 10.6 MJ day^{-1} 1
    (ii) BMR = EAR ÷ PAL / BMR = 10.6 ÷ 1.4; BMR = 7.6;
     [allow CE from (b)(i)]   2

Total 10 marks

Unit 1 Topic 2

1 (a) (i) E: phosphate;
     F: ribose;
     G: uracil;   3
    (ii) nucleus;   1

  (b) (i) Tyr, Val, Glu, Arg;   2
    (ii) translation;   1
    (iii) 1 change in amino acid sequence/primary structure of the protein;
      2 Tyr replaced by a stop codon/UAG;
      3 polypeptide chain/protein would be shorter/eq;
      4 protein would have a different shape/ structure/fold differently/eq;
      5 protein would not function (normally);
      6 RNA polymerase not functioning would mean that no/less transcription could take place;
      7 no/less RNA could be synthesised by this cell;
      8 no/fewer proteins could be made by this cell;   5

Total 12 marks

2 (a) (i) X: phospholipid;
     Y: (channel) protein;
     Z: glycoprotein / carbohydrate;   3
    (ii) P: (passive) diffusion;
     Q: active transport;
     R: facilitated diffusion;   3

  (b) 1 CFTR protein defective/eq;
    2 chloride ions remain in cells/eq;
    3 mucus lacks water/is very sticky/eq;
    4 mucus blocks pancreatic (duct);
    5 fewer enzymes / correctly named enzyme released into the small intestine;
    6 lower concentration of enzymes / fewer active sites;
    7 fewer collisions between substrate and the active site of the enzymes/named substrate and enzyme;   4

Total 10 marks

3 (a) mother's and father's genotypes correctly stated,
e.g. mother **Aa** and father **aa**;
gametes shown clearly and correctly;
F1 genotypes clearly linked to phenotypes;
probability 50%/0.5/1 in 2/½/eq;   3

(b) amniocentesis/chorionic villus sampling/CVS/
pre-implantation genetic diagnosis/PIGD;   1

Total 4 marks

4 (a) 1 reference to vector/eq;
2 reference to functional gene codes for CFTR
protein;
3 CFTR protein in cell (membrane);
4 allows chloride ions to leave the cell;
5 sodium ions (diffuse) out of cells;
6 lowers water potential in the {lumen/airways};
7 draws water out of the cells by osmosis;
8 mucus is kept runny/eq;   4

(b) ¾ /3 out of 4/3 in 4/75%/0.75;   1

(c) amniocentesis/chorionic villus sampling/genetic
screening/IVF/embryo screening/genetic
testing;   1

Total 6 marks

Unit 2 Topic 3

1 (a) (i) true/membrane-bound nucleus;   1
(ii) tissue;
because it is made up of one cell type;
with a function;   2

(b) Spindle fibres drawn from left-hand side to right-
hand side of cell;
(Some) spindle fibres go through the centre of
each chromosome;
No centrioles drawn;   2

cellulose cell wall

cell surface membrane

cytoplasm

(c) cytokinesis;   1

(d) Keep the diameter/initial length of the fibres the
same (from the two different species);
One other appropriate variable described, e.g.
fibres from the same area of the two plant
species;
{Add masses until fibre breaks/ref. to use of force
meter/ref. to fibre stretching};
Suitable safety point made, e.g. cushion to stop
masses landing on the investigators foot;
Ref to comparison, e.g. the one that breaks
second has the higher tensile strength;
Take several readings and find the mean for each
species;   4

Total 10 marks

2 (a) B;   1

(b) 32;
Correct working e.g. 120/24 = 5 (cell
divisions)/{2^5/eq};   2

(c) Different stimuli/named stimulus;
(causes) different genes to be activated/
deactivated (in the two stem cells);
active genes make protein;
via mRNA/eq;
The protein controls different processes in the
two cells;
leading to different cell structures;   3

Total 6 marks

3 (a)   4

Description of phenotype	Genotype only	Environment only	Genotype and environment
a set of non-identical twins being a boy and a girl	✔		
an arctic fox producing dark fur in summer and white fur in winter			✔
a human's lungs enlarge if they do a lot of swimming		✔	
identical twins having different heights at the same age		✔	

  (b) (i)   Many/eq genes;   1
     (ii)   D   1

                                   Total 6 marks

4 (a) (i)   Both substances A and B reduced mean pollen tube growth (in 2 hours)/rate of mean pollen tube growth;
               Substance B caused a greater reduction in mean pollen tube growth;
               Credit correct manipulation of the data, e.g. mean pollen tube length in the presence of substance B reduced the growth by 150 μm;
               (3 answers needed to score max. mark)   3
     (ii)   same {concentration/eq}of substance A and B; same temperature; pollen to have come from the same plant;   1

  (b) (i)   acrosome;   1
     (ii)   {Special vesicles move/fuse with egg membrane/egg cell releases/eq material};
               {Jelly layer/zona pellucida} {thickens/becomes harder};   2

                                     Total 7 marks

Unit 2 Topic 4

1 (a) (i)   B;   1
     (ii)   B;   1

  (b)   4

Statement	Statement correct (✔) or incorrect (✗)
hemicellulose molecules present	✔
calcium is a component of the cellulose molecule	✗
all plant cells have a primary cell wall	✔
a tonoplast may be present	✗

  (c) (i)   cytoplasm/ER filled;
               channels from one cell to the next;
               where no (primary) cell wall is present;
               (and are) lined with membrane;   2
     (ii)   Only pits have a primary cell wall present;
               No ER present in pit;
               Pits do not extend from one cell to the next;   2

                               Total 10 marks

2   There was size variation within the bird population;
      Bird size is under the influence of {alleles/eq};
      The birds with the allele for larger size were selected;
      (and) they survived the cold/were better insulated;
      so were able to breed and pass on the allele for larger size; so the offspring had the allele for larger size, the offspring were larger;   4

                               Total 4 marks

3 (a) (i)   to allow comparison with other solution;
               so any difference in length increase must be due to the nitrates;   1
     (ii)   to reduce genetic variation;   1
    (iii)   difference in length found/final length minus original length (for each seedling);
               all 10 added together;
               (and then) divided by 10/eq;   2
    (iv)   Nitrates increased the length over 14 days more than without nitrates;
               Correct manipulation of the data e.g. a 100% increase compared to situation without nitrates;   2

  (b)   nitrate;
         nitrogen;
         {protein/named protein/polypeptide};
         middle;
         magnesium;   5

(c) {reactant/ eq} in photosynthesis;
major component of cytoplasm;
medium for transport of mineral ions;
involved in turgor in cells;
solvent for chemical reactions;
any other sensible use;   4

Total 15 marks

4 (a)   3

Description	Phase
A few patients are involved.	2;
A few healthy volunteers are involved.	1;
A double-blind trail is undertaken.	3;

(b) 10 arbitrary units;
Difference between apparent speed of recovery
for drug and the speed of recovery induced by
the placebo;   2

Total 5 marks

5 (a) Site 2;
(because) has greatest number of different
species/10 species;
(and) species richness is a measure of the number
of different species present in an area;   2
(b) Idea that sample is not representative;   1
(c) Site 3;   1

Total 4 marks

Answers to specimen paper questions

Unit 1

1 A platelets/thrombocytes;
  B prothrombin;
  C enzyme;
  D fibrinogen;
  E fibrin;
  F cells/erythrocytes/platelets/thrombocytes; 6

Total 6 marks

2 (a) 1 translation;
    2 transcription;
    3 translation;
    4 translation;
    5 transcription; 5

  (b) (i) glutamine 1
    (ii) cysteine glutamine cysteine arginine proline
        proline; 1
    (iii) ATC; 1
    (iv) U G U G A A U G U C G G C C A C C C; 1
    (v) The polypeptide chain would be no more
       than 89 amino acids long; 1

Total 10 marks

3 (a) 1 {fatty acids/tails} are {hydrophobic/non-polar};
    2 so orientate themselves away from {water/polar
      environment}/eq;
    3 {phosphate/heads} are {hydrophilic/polar};
    4 so can interact with {water/polar environment}/
      eq;
    5 reference to {cytoplasm/tissue fluid/eq} as the
      polar environment; 3

  (b) (i) Any two from:
      temperature,
      surface area/volume (of beetroot),
      part,
      age,
      variety,
      storage,
      source,
      volume of ethanol,
      same {wavelength/filter}; 2
    (ii) 1 {cells/membranes/eq} damaged (by
        cutting up of pieces)/eq;
      2 (as a result pigment) could leak out of
        {vacuoles/cells}; 2
    (iii) rinse pieces (thoroughly)/dab pieces dry/eq; 1

  (c) (i) increased ethanol concentrations, increases
      intensity/eq; 1
    (ii) 1 reference to {disruption/eq} of membrane;
      2 ethanol is a (non–polar/organic) solvent;
      3 dea that {lipids/eq} dissolve (in alcohol);
      4 idea that increase in ethanol causes
        solution to be less polar;

    5 idea that orientation of phospholipids
      depends on water around it; 2

Total 11 marks

4 (a) 1 thick wall drawn;
    2 {two/three/four} layers indicated;
      (Max. two from the following correctly
      labelled:)
    3 lumen;
    4 {endothelium/epithelium/endothelial layer/
      epithelial layer/tunica intima};
    5 {(smooth) muscle/elastic fibres/elastin/tunica
      media };
    6 {connective tissue/tunica adventitia}; 3

  (b) 1 idea of {wide wall/eq} (to withstand) blood
      under high pressure;
    2 reference to narrow lumen to maintain high
      pressure;
    3 reference to presence of {elastic fibres/eq} to
      allow vessel to stretch;
    4 recoil {maintains pressure/squeezes blood};
    5 reference to (smooth) muscle contracts to
      {squeeze/eq} blood along};
    6 idea that {smooth lining/eq} reduces friction;
    7 {folded lining/eq} to allow artery to stretch/eq;2

  (c) 1 (walls of) veins more than one layer of cells and
      capillaries one layer /eq;
    2 (walls of) veins contain {connective tissue
      /(smooth) muscle/collagen/elastic tissue},
      capillaries do not/eq;
    3 veins have valves in them and capillaries do
      not/eq;
    4 veins do not have pores but capillaries do/eq;
    5 veins have wide lumen, capillaries have narrow
      lumen /eq; 2

Total 7 marks

5 (a) (i) 1 A has a {greater/eq} effect than B/eq;
      2 A lowers total cholesterol more than B/eq;
      3 A lowers LDL more than B/eq;
      4 A raises HDL more than B /eq;
      5 manipulation of figures to quantify mp 2
        or 3 or 4; 3
    (ii) 1 drug A;
      2 the {total cholesterol/LDL} levels are lower;
      3 statins inhibit cholesterol synthesis;
      4 statins result in more LDL receptors on liver
        cells;
      5 so more LDL will be {cleared /eq} from the
        blood /eq; 3

  (b) Any two from:
    gastrointestinal {problems/cancer} e.g.
    constipation, bowel complaints,
    {joint/muscle} problems, e.g. cramps, myositis,
    pain, myopathy,
    muscle breakdown,

liver problems,
kidney problems,
mental health problems, e.g. depression,
reduced vitamin uptake,
respiratory cancer; 2

(c) (i) 1 reference to the (general) increase in heart
disease with age;
2 more 18–44 year old females develop heart
disease than males/eq;
3 in all other age groups more males have
heart disease than females/eq;
4 greatest difference between females and
males in the group 65–74;
5 credit manipulation of figures; 3

(ii) 1 {420/425} – {30/35}/390/385/395;
2 11 – 13; 2

Total 13 marks

6 (a) B 1
(b) D 1
(c) B 1
(d) B 1

Total 4 marks

Unit 2

1 (a) amino acids;
endoplasmic;
vesicles;
modified/eq; 4

(b) (i) 1 = DNA;
2 = matrix (of mitochondrion);
3 = crista; 3
(ii) made up of 2/inner and outer;
membranes; 2

(c) rER ribosomes are attached to a membrane unlike
those in mitochondria/{mitochondrial ribosomes
smaller/converse}; 1

Total 10 marks

2 (a) (i) C; 1
(ii) B; 1

(b) (i) down style/eq;
around ovary;
to micropyle; 2
(ii) Both increase for first 18 hours;
Fastest increase in length in the first 6 hours;
After storage always smaller length than
immediately after collection/converse;
After storage stops growing after 18 hours/
only immediately after collection pollen
continues to grow after 18 hours;
Credit manipulation of the data; 3

Total 7 marks

3 (a) 3

Polysaccharide	Chemical composition	Shape
starch/eq;		may be branched/ {coiled/eq};
	made up of many β glucose;	

(b) (i) {many/eq}cellulose molecules;
all lying parallel to each other;
joined by {hydrogen/H} bonds; 2
(ii) C; 1

(c) (i) Mean mass needed to break fibre increases
as diameter of fibre increases;
linear increase between 0.2 mm and 0.5 mm;
Credit manipulation of the data; 2
(ii) Smaller error bar at 0.2 mm/converse;
All fibres broken between 400 and 600 g
/converse;
Idea of spread of data is smaller at 0.2 mm,
therefore more data around the mean; 2
(iii) All fibres broken at {1400 g/same mass/eq}; 1
(iv) (Manmade fibre came) from oil;
(and) oil is a fossil fuel;
(which is) non-renewable/replaced more
slowly (naturally) than being used (by man); 2

Total 13 marks

4 (a) Archaea;
{Eukaryota/eq}; 2

(b)

Statement	Letter
site where biochemical reactions take place	
contains genetic material, including genes for antibiotic resistance	C
structure made of peptidoglycan that is only found in bacteria	A
membrane that regulates what enters and leaves the bacterium	B
structure used for movement	D

All 4 correct for 3 marks, 3 correct for 2 marks &
2 correct for 1 mark; 3

(c) In context:
{Genetic variation/eq} within bacteria;
Some bacteria had {allele for antibiotic resistance/
advantageous allele};
{presence of antibiotic/change in conditions/eq};
Only allowed those with allele to survive;
Breed and pass on allele for antibiotic resistance
to offspring;
Over many generations, all bacteria in colony
would have this advantageous allele;
Ref. to change in allele frequency; 5

(d) (i) If unsafe then less likely to cause serious harm {as volunteers are well/ref to effectiveness of body defence system};   1

(ii) To compare the response with the placebo to response with the drug;
The difference being the actual effect of the drug;
To remove any non-clinical response;   2

Total 13 marks

5 (a) Stop individuals repeatedly breeding with same partners;
Use a stud book/eq to effectively;
Select partner;
Use sperm/egg from an individual from another zoo;
Swap animals between zoos;   2

(b) Explain:
captive breeding programmes/reintroduction programmes/importance of biodiversity/ importance of conservation /any other sensible answer;   1

(c) (i) (increasing storage time) decreases percentage germination success;
Most rapid decline in first 60 days;
No germination success/eq after 120 days;
Credit manipulation of the data;   3

(ii) Collect new seeds (to replace old ones);
Allow seeds to germinate, grow and mature, then collect seeds from these plants;
Keep the current seeds at a {temperature below room temperature/given temp};
{Remove water from them/dehydrate them/ dry them/eq};   2

Total 8 marks

6 (a) embryos from {IVF/eq};
Idea of being embryos that would not be used/ implanted in woman;   2

(b) (i) The research may lead to treatments for various conditions/named condition;
The spare embryos would be discarded anyway;
The embryo should not be considered as human at such an early stage;
Idea of more cell types produced from embryonic stem cells than other stem cell types;   2

(ii) Embryos are (potential) {humans/babies/have a right to life/eq};
from {conception/eq};
ethical/religious objections;
Idea that women having IVF will be 'pressured' into producing surplus embryos;
Idea that could get into the 'wrong hands';
Idea of designer babies;   2

Total 6 marks

7 (a) Idea of the amount of genetic variation;
within a species;
Detail of genetic variation such as the number of alleles at a gene locus;   2

(b) {As the ratio of males to females increases, the genetic diversity increases / positive correlation};
in a non-linear manner/correct description of trend;   2

(c) (with few males) breeding;
fewer different alleles/eq;
are being passed on by the male;
(so) low genetic diversity;   2

Total 6 marks